Intermediate and CLEP Algebra

By Vali Nasser

Copyright © 2016

E-book editions are also available for this title. For more information email:

valinasser@gmail.com

All rights reserved by the author. No part of this publication can be reproduced, stored in a retrieval system, or transmitted in any form or by any means, electronic, mechanical, photocopying, recording or otherwise, without the prior permission of the publisher and/or author.

Sept 2016

ISBN-13: 978-1537391502

ISBN-10: 153739150X

Every effort has been made by the author to ensure that the material in this book is up to date and in line with the requirements to pass 'Intermediate and CLEP Algebra' at the time of publication. The author will also do his best to review, revise and update this material periodically as necessary. However, neither the author nor the publisher can accept responsibility for loss or damage resulting from the material in this book

INTRODUCTION 9
- 25% Algebraic operations 9
- 25% Equations and inequalities 9
- 30% Functions and their properties* 10
- 20% Number systems and operations 10

NUMBER SECTION 1 11
Rational and Irrational Numbers 11

IMAGINARY NUMBERS: 15
An imaginary number gives a negative result when 15
Complex Numbers 15
Examples: 15

SCIENTIFIC NOTATION 16

NUMBER WORK SECTION 1: PRACTICE QUESTIONS 18

ANSWERS TO NUMBER WORK SECTION 1: PRACTICE QUESTIONS 19

NUMBER SECTION 2 20
Exponents or Indices 20
Fractional Exponents 21

More on Rational and Irrational Numbers 21

MORE RADICALS 23

NUMBER WORK SECTION 2: PRACTICE QUESTIONS 24

ANSWERS TO NUMBER WORK SECTION 2: PRACTICE QUESTIONS 25

ALGEBRA SECTION 1: 26

Basic Algebra 26

Multiplying positive and negative numbers. 27

Dividing positive and negative numbers. 27

Simplifying algebraic expressions 28

Simplifying algebraic expressions 31

Multiplying out brackets. 32

ALGEBRA SECTION 1: PRACTICE QUESTIONS 39

Simplify the following: 39

ANSWERS TO SIMPLIFYING EXPRESSIONS 40

ALGEBRA SECTION 2 41

Factoring 41

ALGEBRA SECTION 2: PRACTICE QUESTIONS 42

ANSWERS TO FACTORING EXPRESSIONS: 43

ALGEBRA SECTION 3 44

Algebraic Substitution and Formula 44

FORMULA 45

Changing the subject of a formula 46

ALGEBRA SECTION 3: PRACTICE QUESTIONS 49

Practice Questions on Change the Subject of formula 49

ANSWERS TO CHANGE THE SUBJECT OF FORMULA: 50

ALGEBRA SECTION 4 51

Solving equations 51

Word Problems using Algebra 54

ALGEBRA SECTION 5 57

Algebraic Proofs 57

ALGEBRA SECTION 5: PRACTICE QUESTIONS 58

Practice Questions on Algebraic Proofs: 58

ANSWERS TO ALGEBRAIC PROOFS: 59

ALGEBRA SECTION 6 61

Simultaneous Equations 61

PRACTICE QUESTIONS ALGEBRA SECTIONS 5 AND 6 65

ANSWERS TO ALGEBRA SECTIONS 5 AND 6 67

ALGEBRA SECTION 7 68

Solving Quadratic Equations 68

SOLVING QUADRATIC INEQUALITIES 74

GRAPHS OF QUADRATIC EQUATIONS 77

CUBIC EQUATION 81

EXPONENTIAL GRAPHS 82

SOLVING EQUATIONS USING GRAPHICAL METHODS 83

ALGEBRA SECTION 8 86

Equation of a Circle 86

PRACTICE QUESTIONS ON EQUATION OF A CIRCLE 89

ANSWERS TO EQUATION OF CIRCLE QUESTIONS 90

ALGEBRA SECTION 9 92

Remainder and Factor Theorem 92

Remainder Theorem 92

Factor Theorem 92

SOLVING CUBIC EQUATIONS 94

PRACTICE QUESTIONS ALGEBRA SECTION 9 97

ANSWERS TO ALGEBRA SECTION 9 98

ALGEBRA SECTION 10 101

Linear equations 101

USING RATIOS TO FIND CO-ORDINATES 105

WORKING OUT EQUATIONS OF 'NORMALS' AND 'PARALLEL' LINES 106

Finding parallel lines 107

ALGEBRA SECTION 11 108

Logarithms 108

Some useful definitions when using logarithms 108

A logarithm (log) is the inverse of exponentiation 108

Notice that logarithm is an inverse function of an exponential function 108

Inverse logarithm calculation 108

LOGARITHMIC FUNCTION 109

Important logarithm rules to remember: 109

Some more examples 111

PRACTICE QUESTIONS ON LOGARITHMS 114

ANSWERS TO LOGARITHM QUESTIONS 115

ALGEBRA SECTION 12 116

Sequences 116

QUADRATIC SEQUENCE 118

RECURRENCE RELATIONS.. 120

FINDING THE SUM OF NATURAL NUMBERS: 123

GEOMETRIC SEQUENCES AND SUMS: 125

Example 1:..125

Summing a Geometric Series 126

PRACTICE QUESTIONS ON SEQUENCES 128

ANSWERS TO QUESTIONS ON SEQUENCES................................ 129

ALGEBRA SECTION 13 130

Sets and Venn Diagrams 130

Symbols and notation associated with Sets: 130

ALGEBRA SECTION 14 145

Binomial expansion 145

BINOMIAL THEOREM.. 147

Probability re-visited 147

Binomial Probability Function 150

ALGEBRA SECTION 15 154

Matrices 154

A matrix is simply an array of numbers 154

MULTIPLYING A MATRIX BY ANOTHER MATRIX 156

DETERMINANT OF A MATRIX 162

What is it used for? 162

Symbol 162

Calculating the Determinant... 162

PRACTICE QUESTIONS ON MATRICES 165

ANSWERS TO PRACTICE QUESTIONS ON MATRICES 167

Introduction

This book on **'Intermediate and CLEP Algebra'** has many examples as well as practice questions that will help you get up to speed with the algebra required to CLEP level. **Although it starts gently with number work this quickly builds up to rigorous algebra to intermediate and CLEP level.**

If you are going to take the CLEP exam in Algebra you will probably know that the type of questions given are multiple choice type questions. However you will find the practice questions in this book expect you to give a single answer. This, I believe, will help you to work out all the steps necessary and avoid guessing. You will then be much more confident when doing the actual exam.

The assessment on the appropriate topics in CLEP algebra is approximately as shown below:

25%
Algebraic operations

- Factoring and expanding polynomials
- Operations with algebraic expressions
- Operations with exponents
- Properties of logarithms

25%
Equations and inequalities

- Linear equations and inequalities
- Quadratic equations and inequalities
- Absolute value equations and inequalities
- Systems of equations and inequalities
- Exponential and logarithmic equations

30%
Functions and their properties*

- Definition and interpretation
- Representation/modeling (graphical, numerical, symbolic, and verbal representations of functions)
- Domain and range
- Algebra of functions
- Graphs and their properties (including intercepts, symmetry, and transformations)
- Inverse functions

20%
Number systems and operations

- Real numbers
- Complex numbers
- Sequences and series
- Factorials and Binomial Theorem
- Determinants of 2-by-2 matrices

About the Author
The author of this book has experience in both consultancy work and teaching. As a specialist mathematics teacher he has tutored and taught mathematics in schools as well as in adult education. The Author has worked in the USA as well as the UK. The author's initial book 'Speed Mathematics Using the Vedic System' has a significant following and has been translated into Japanese and Chinese as well as German. His book 'Pass the QTS Numeracy Skills Test with Ease' has been very popular with teacher trainees in the UK. He is also the author of several Math books. He hopes that his new book **Intermediate and CLEP Algebra** will be helpful to pupils who need to revise the appropriate topics for this particular exam.

Number Section 1

Natural numbers: These are: {1, 2, 3, ...}

Whole numbers: The numbers {0, 1, 2, 3, ...} are whole numbers

Integers: Positive and <u>negative</u> whole numbers, as well as zero: {..., -2, -1, 0, 1, 2,...}.

Rational and Irrational Numbers

Numbers can be either rational or irrational

Any number that can be written as p/q is a rational number, where p and q are whole numbers and q is not zero. Basically, the number is well defined and we know or can predict its pattern. Whole Numbers, Integers as well as fractions such as those shown above and below are rational.

Examples of rational numbers are: $5 = \frac{5}{1}$, $-2 = \frac{-2}{1}$, $\frac{1}{2} = 0.5$, $\frac{2}{5} = 0.4$, $\frac{1}{3} = 0.33333$ (recurring)

$\frac{0}{5} = 0$, $\frac{4}{33} = 0.1212121212.....$

Examples of irrational numbers are: π, $\sqrt{2}$, $\sqrt{3}$ or $5\sqrt{7}$

For square roots and cube roots those with perfect roots are rational whereas others are irrational. So for example $\sqrt{25} = 5$ is rational, $\sqrt[3]{27} = 3$ is rational but as we saw earlier **$\sqrt{2}$ is irrational.**

For example π or √2 <u>do not have a predictable pattern</u>. We can approximate them but not calculate them exactly.

Factors: A factor is a number that divides exactly into another number for example, the number 2 in the case of even numbers.

3 is a factor of 9, as 3 goes exactly into 9. Other factors of 9 are 1 and 9.

15, has two factors other than 15 and 1. The two factors are 5 and 3, since both these numbers go exactly into 15. **Example:** Find all the factors of 21. The factors are: 1, 3, 7 and 21 (since all these numbers divide exactly into 21)

Prime numbers: A prime number is a natural number that can be divided only by itself and by 1 (without a remainder). For example, 11 can be divided only by 1 and by 11. Prime numbers are whole numbers greater than 1. So for example the first 10 prime numbers are: 2, 3, 5, 7, 11, 13, 17, 19, 23 and 29. **Be careful that an odd number is not necessarily a prime number.** For example **9 is not a prime number** as its factors are 1, 3 and 9 and **prime numbers should have only two factors, 1 and the number itself. Also, note that 2 is a prime number, the only even number that can be divided by 1 and itself!**

Square numbers and square roots

Squaring a number is simply multiplying a number by itself.

So 4^2 means $4 \times 4 = 16$, 12^2 means $12 \times 12 = 144$ and so on.

The square root is written like this $\sqrt{}$ and means finding a number which when multiplied by itself gives you the number inside the square root. The number under the $\sqrt{}$ (**radical sign**) is called a **radicand**.

So, in $\sqrt{7}$ the number 7 under the radical sign is called a radicand.

Example: Find $\sqrt{16}$. The answer is clearly 4. Since $4 \times 4 = 16$

Let us consider some other square roots.

$\sqrt{49} = 7$, $\sqrt{121} = 11$, $\sqrt{100} = 10$, $\sqrt{225} = 15$,

$\sqrt{256} = 16$, $\sqrt{324} = 18$

Cubes and cubic roots

Cubing a number is simply multiplying the number by itself three consecutive times. A cube of a number is written as x^3, where x is the number.

So, for example, 5^3 means $5 \times 5 \times 5 = 25 \times 5 = 125$

Similarly, $6^3 = 6 \times 6 \times 6 = 216$, $7^3 = 7 \times 7 \times 7 = 343$, $9^3 = 9 \times 9 \times 9 = 729$,

$10^3 = 10 \times 10 \times 10 = 1000$

Cube Roots

Cube roots are found by finding a number which when cubed gives you the number inside the cube root.

So for example the cube root of 125 is written as $\sqrt[3]{125}$

Also we know that 5×5×5 =125, so that $\sqrt[3]{125} = 5$

Imaginary Numbers:

An imaginary number gives a negative result when squared

For example consider the square root of -1

$i = \sqrt{-1}$

Since when we square i we get -1

$i^2 = -1$

Examples of Imaginary Numbers:

    4i,   2.5i, $-2.7i$, etc

Complex Numbers

A Complex Number is a combination of a Real Number and an Imaginary Number:

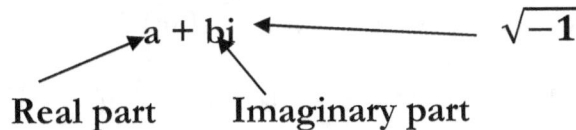

Real part   Imaginary part

Examples:

    $2 + i$ $45 + 3i$ $0.1 - 4.2i$ $\sqrt{3} + i/2$

Basically, a complex Number is just **two types of numbers added together** (a **Real** and an **Imaginary** Number).

Scientific Notation

Scientific notation helps us to write **very large** or **very small** numbers in a more elegant way.

Example 1: Write three million in scientific notation.

Three million = 3000,000 = 3 × 1000,000

Now 3 × 1000,000 = 3 × 10^6

3 × 10^6 is the required scientific notation.

Example 2: Write 4000,000 in scientific notation.

We know that 4000,000 = 4 × 1000,000

Hence we can write 4 × 1000,000 as 4 × 10^6. This is in scientific notation.

Example 3: Change 4500,000 to scientific notation.

We know that 4500,000 = 4.5 × 1000,000

So 4.5 × 1000,000 = 4.5 × 10^6. This is in scientific notation..

Now consider very small numbers:

Example 4: Write 0.0004 in scientific notation.

We can re-write 0.0004 as 4 ÷ 10,000

Now, 4 ÷ 10,000 can be written as $\dfrac{4}{10 \times 10 \times 10 \times 10} = 4 \times 10^{-4}$

Notice, $\frac{1}{10} = 10^{-1}$, $\frac{1}{100} = 10^{-2}$, $\frac{1}{1000} = 10^{-3}$ and $\frac{1}{10000} = 10^{-4}$

Summary:

To convert a number to scientific notation change the number to the form a × 10^n where 1 ≤ a < 10 (a is between 1 and 10) and n is a whole positive or negative number)

Number Work Section 1: Practice Questions

(1) Find the cubic root of 729

(2) Simplify $\sqrt{49} \times \sqrt{121}$

(3) What sort of number is 4 + 3i?

(4) What is the square of an imaginary number $\sqrt{-1}$

(5) Given that 8km is approximately 5miles. How many kilometres are there in 75 miles?

(6) In a class of 28 pupils, 8 pupils have extra math tuition. What is the proportion of pupils in this class that do not have extra math tuition?

(7) $280 is divided in the ratio 5: 2: 1. Find the largest part.

Answers to Number Work Section 1: Practice Questions

(1) Answer: 9

(2) Answer: 77

(3) Answer: complex

(4) Answer: -1

(5) Answer: 120 km

(6) Answer: $\dfrac{5}{7}$

(7) Answer: The largest part is $175

Number Section 2

Exponents or Indices

You are probably already familiar with squares, square roots, cubes and cube roots. Exponents, powers, Indices/Index Numbers are simply the power by which a base number is raised. So just as 4^3 (4 cubed) means 4 to the power of 3, these 'powers' as mentioned earlier are also referred to as exponents or indices. So 5^6 simply means 5 raised to the power of six. So in this case 5^6 means 5×5 5×5×5×5! (5 is called the base number and 6 is the power or exponent) It is interesting to note that if you multiply two or more same base numbers with exponents for example: $5^6 \times 5^3$ you simply add the indices/exponents to get 5^9 (5 to the power 9). Reason: 5^6 means 5×5×5×5×5×5 and 5^3 means 5×5×5 so $5^6 \times 5^3$ = (5×5×5×5×5×5) × (5×5×5) = 5^9

Similarly, for division, you simply subtract the exponents. Consider $5^6 \div 5^3$. This means we need to work out $\frac{5 \times 5 \times 5 \times 5 \times 5 \times 5}{5 \times 5 \times 5}$ which cancels down to 5×5×5 or 5^3 So you can see that when dividing the **same** base numbers with indices you simply subtract the exponents.

The examples below will help you to consolidate the manipulation of the same base numbers with exponents.

Example 1: $7^8 \times 7^4 \times 7^6 = 7^{18}$ (simply add the exponents 8 + 4 +6 =18, hence the answer is: 7^{18})

Example 2: $9^{12} \div 9^5 = 9^7$ (simply subtract 5 from 12 to get 7, hence the answer is: 9^7)

Finally, you can also have negative exponents which are inverses of the base numbers with the appropriate indices.

Example 1: $5^{-1} = \frac{1}{5}$ (Also called the reciprocal of 5).

Example 2: $6^{-2} = \frac{1}{6^2}$

Example 3: $5^{-6} = \frac{1}{5^6}$

Fractional Exponents: Examples: (i) $2^{1/2}$ (2 to the power of $\frac{1}{2}$), (ii) $27^{1/3}$ (27 to the power of $\frac{1}{3}$. (It's worth noting that $2^{1/2}$ is the same as $\sqrt{2}$, $27^{1/3}$ is the same as $\sqrt[3]{3}$ and $8^{2/3}$ means (8 to the power of $\frac{2}{3}$.)

More on Rational and Irrational Numbers

As we saw earlier numbers can be either rational or irrational

Any number that can be written as p/q is a rational number, where p and q are whole numbers and q is not zero. Basically, the number is well defined and we know or can predict its pattern.

Examples of rational numbers are: $5 = \frac{5}{1}$, $-2 = \frac{-2}{1}$, $\frac{1}{2} = 0.5$, $\frac{2}{5} = 0.4$, $\frac{1}{3} = 0.33333$ (recurring)

$\frac{0}{5} = 0$, $\frac{4}{33} = 0.1212121212.....$

Examples of irrational numbers are: $\pi, \sqrt{2}, \sqrt{3}$ or $5\sqrt{7}$

For square roots and cube roots those with perfect roots are rational whereas others are irrational. So for example $\sqrt{25} = 5$ is rational, $\sqrt[3]{27} = 3$ is rational but as we saw earlier $\sqrt{2}$ is irrational.

For example π or $\sqrt{2}$ do not have a predictable pattern. We can approximate them but not calculate them exactly.

Summary for Exponential expressions:

Rules:

(1) $a^m \times a^n = a^{m+n}$

(2) $a^m \div a^n = a^{m-n}$

(3) $(a^m)^n = a^{m \times n}$

(4) $a^0 = 1$

(5) $a^{-1} = \dfrac{1}{a}$

(6) $a^{-m} = \dfrac{1}{a^m}$

More Radicals

Sometimes radicals are expressions with irrational square roots. There are some useful rules associated with them.

1: $\sqrt{2} \times \sqrt{2} = \sqrt{4} = 2$

2: $\sqrt{3} \times \sqrt{2} = \sqrt{6}$

3: $\dfrac{\sqrt{6}}{\sqrt{2}} = \sqrt{\dfrac{6}{2}} = \sqrt{3}$

4: $(\sqrt{p} + \sqrt{q})^2 = (\sqrt{p} + \sqrt{q}) \times (\sqrt{p} + \sqrt{q}) = p + 2\sqrt{pq} + q$

5. $(p + \sqrt{q})(p - \sqrt{q}) = p^2 + p\sqrt{q} - p\sqrt{q} - q = p^2 - q$

6. $\dfrac{2}{\sqrt{3}}$ (Multiply top and bottom by $\sqrt{3}$) so we have: $\dfrac{2}{\sqrt{3}} \times \dfrac{\sqrt{3}}{\sqrt{3}} = \dfrac{2\sqrt{3}}{3}$

(we call this rationalising the denominator)

7. $\dfrac{2}{1-\sqrt{3}}$ to simplify this we need to <u>rationalise</u> the denominator.

To do this we simply multiply top and bottom by $(1 + \sqrt{3})$

So we have, $\dfrac{2}{1-\sqrt{3}} = \dfrac{2}{1-\sqrt{3}} \times \dfrac{1+\sqrt{3}}{1+\sqrt{3}} = \dfrac{2(1+\sqrt{3})}{1-3} = \dfrac{2(1+\sqrt{3})}{-2} = -(1+\sqrt{3})$

Number Work Section 2: Practice Questions

1 Simplify the following

(a) $2^3 \times 2^4$ (b) $p^6 \div p^7$ (c) $(a^m \times a^n) \div a^k$ (d) $2^{\frac{1}{2}} \times 2^{\frac{-3}{2}}$

2 Write down the following square and cubic roots in its simplest form

(a) $\dfrac{1}{\sqrt{64}}$ (b) $\dfrac{\sqrt{4}}{\sqrt{121}}$ (c) $\dfrac{\sqrt{324}}{18}$ (d) $\sqrt[3]{27}$

(e) $\dfrac{36}{\sqrt[3]{729}}$

3 Write $\sqrt{98} - \sqrt{128} + 5\sqrt{72}$ in the form $a\sqrt{2}$

4 Rationalize the following radicals:

(a) $\dfrac{\sqrt{3}}{1-\sqrt{3}}$ (b) $\dfrac{1+\sqrt{2}}{1-\sqrt{2}}$ (c) $\dfrac{5+\sqrt{n}}{5-\sqrt{n}}$

Answers to Number Work Section 2: Practice Questions

1 (a) 2^7 (b) p^{-1} or $\frac{1}{p}$ (c) a^{m+n-k} (d) 2^{-1}

2 (a) $\frac{1}{8}$ (b) $\frac{2}{11}$ (c) 1 (d) 3 (e) 4

3 $29\sqrt{2}$

4 (a) $\frac{-3}{2} - \frac{\sqrt{3}}{2}$ (b) $-3 - 2\sqrt{2}$ (c) $\frac{25+10\sqrt{n}+n}{25-n}$

Algebra Section 1:

Basic Algebra

In algebra we often use letters instead of numbers. There are some basic conventions and rules of algebra that you should be familiar with to progress in this subject. This chapter will be useful for you if you have forgotten your algebra.

If you see	we mean
$x = y$	x equals y
$x > y$	x is greater than y
$x < y$	x is less than y
$x \geq y$	x is greater than or equal to y
$x \leq y$	x is less than or equal to y
$x + y$	the sum of x and y
$x - y$	subtract y from x
xy	x times y
x/y	x divided by y
$x \div y$	x divided by y
x^n	x to the power n
$x(x + y)$	x times the sum of x + y

Also note that:

$x(x+y) = x^2 + xy$

$x^2(x + x^2 + y) = x^3 + x^4 + x^2y$

In general, a x a x a x a ....(n times) $= a^n$

You also need to know these algebraic rules for the multiplication and division of positive and negative numbers.

Multiplying positive and negative numbers.

$(+) \times (+) = +$ (a plus number times a plus number gives us a plus number)

$(+) \times (-) = -$ (a plus number times a minus number gives us a minus number)

$(-) \times (+) = -$ (a minus number times a plus number gives us a minus number)

$(-) \times (-) = +$ (a minus number times a minus number gives us a plus number)

Dividing positive and negative numbers.

$(+) \div (+) = +$ (a plus number divided by a plus number gives us a plus number)

$(+) \div (-) = -$ (a plus number divided by a minus number gives us a minus number)

$(-) \div (+) = -$ (a minus number divided by a plus number gives us a minus number)

$(-) \div (-) = +$ (a minus number divided by a minus number gives us a plus number)

Summary: For both multiplication and division, like signs gives us a plus sign and unlike signs gives a minus sign

Also when adding and subtracting it is worth knowing that:

When you add two minus numbers you get a bigger minus number.

Example 1: $-4 - 6 = -10$

When you add a plus number and a minus number you get the sign corresponding to the bigger number as shown below:

Example 2: $+6 - 9 = -3$, whereas, $-6+9 = 3$

When you subtract a minus from a plus or minus number you need to note the results as shown below:

Example 3: $6 - (-3)$ we get $6+3 = 9$ (since $-(-3) = +3$)

Example 4: $7 - (+3)$ we get $7 - 3 = 4$ (since $-(+3) = -3$)

In this case note that $-(-) = +$. Also, $+(-) = -$ and $-(+) = -$.

Simplifying algebraic expressions

Example 1: Simplify $3x + 4x + 5x$

Method: We simple add up all the x's.

Hence we get $3x+4x+5x = 12x$

Example 2: Simplify $3x + 4x + 3y + 5y$

Method: Add up all the like terms.

So we get $3x + 4x + 3y + 5y = 7x + 8y$

(Notice we add up all the x's and then all the y's)

Example 3: Simplify $3m + 4y + 2m - 3y$

Method: as before, we add and subtract like terms.

Now $3m + 2m = 5m$ and $4y - 3y = 1y$ or just y.

So we can write $3m + 4y + 2m - 3y = 5m + y$.

Multiplying out brackets.

Example 1: Expand and simplify $3(2x + 5) + 4(2x + 7)$

Method: Multiply 3 by each term in the first bracket then 4 by each term in the second bracket. The final step is to simplify by collecting up the like terms.

$3(2x+5) + 4(2x+7) = 6x + 15 + 8x + 28 = 14x + 43$

Notice the last step is simply adding $6x + 8x$ and then $15+28$.

$x(x + y) = x^2 + xy$

$x^2(x + x^2 + y) = x^3 + x^4 + x^2 y$

In general, $a \times a \times a \times a \ldots\ldots(n \text{ times}) = a^n$

You also need to know these algebraic rules for the multiplication and division of positive and negative numbers.

Also remember the following terminology of Monomials, Binomials, Trinomials and in general Polynomial expressions:

Examples of Monomials:

x is a monomial expression **(has only one term)**

5 is a monomial expression

7k is a monomial expression

Examples of Binomial expressions (two unlike terms)

z + 3

x + m

3x − 5y

Examples of Trinomial expressions (three unlike terms)

m + n + 3

3x + 3y − 6

5a + 3n + 4

A Polynomial is simply an expression with two or more unlike terms as shown below: (Binomials and Trinomials are also polynomials!)

Examples of polynomials:

m + n + z − 3

4d + 5c – 3e + k

a + b + 3

c – 2d

Summary: <u>For both multiplication and division, like signs give us a plus sign and unlike signs give a minus sign</u>

Also when adding and subtracting it is worth knowing that when **you add two minus numbers you get a bigger minus number.**

Example 1: –4 – 6 = –10

When you add a plus number and a minus number you get the sign corresponding to the bigger number as shown below:

Example 2: +6 – 9 =–3, whereas, –6+9 =3

When you subtract a minus from a plus or minus number you need to note the results as shown below:

Example 3: 6 –(– 3) we get 6+3 =9 (since –(–3) =+3)

Example 4: 7 –(+3) we get 7 – 3 =4 (since –(+3) =–3)

In this case note that – (–) =+. Also, +(–) = – and –(+) = –.

Simplifying algebraic expressions

Example 1: Simplify $3m^2 + 4y^3 + 4m^2 - 5y^3$

Method: We add and subtract like terms.

Now $3m^2 + 4m^2 = 7m^2$ and $4y^3 - 5y^3 = -y^3$

Hence, $3m^2 + 4y^3 + 4m^2 - 5y^3 = 7m^2 - y^3$

Example 3: Simplify $ax^2 \times a^4 x^3$

Method: Multiply the two expressions and add the indices for similar terms : This means $ax^2 \times a^4 x^3 = a^5 x^5$

Example 4 : Simplify $\dfrac{y^3}{x^2} \div \dfrac{y^2}{x}$

Method : Using the rules of indices we get : $\dfrac{y^3}{x^2} \div \dfrac{y^2}{x} = \dfrac{y^3}{x^2} \times \dfrac{x}{y^2}$
$= \dfrac{y}{x}$

(**N.B.** In the above example you can of course cancel down to get the same answer)

Multiplying out brackets.

Example 1: Expand and simplify $3(2x+5) + 4(2x+7)$

Method: Multiply 3 by each term in the first bracket then 4 by each term in the second bracket. The final step is to simplify by collecting up the like terms.

$3(2x+5) + 4(2x+7) = 6x+15+8x+28 = 14x + 43$

Example 2: Work out $(2x+3)(2x+4)$

When we have to multiply out two brackets we have to multiply each term in the first bracket by each term in the second bracket. We then simplify the resulting expression as before. **An easy way to multiply out two brackets is to use the grid method as shown below:**

First put each of the terms of each bracket on the outside grid as shown

×	2x	+3
2x		
+4		

Step2: Multiply each outside term together. So that for example 2x X 2x =$4x^2$. The other results are shown inside the grid.

×	2x	+ 3
2x	$4x^2$	+ 6x
+ 4	8x	+12

After multiplying out the terms, the answer is found by adding all the terms inside the grid and simplifying the resulting expression.

So we have, $4x^2$ + 6x + 8x +12 (These are all the terms inside the grid)

Finally, $4x^2$ + 6x + 8x +12 = $4x^2$ +14x +12

Another example will help consolidate the process:

Multiply out $(2x - 3)(3x + 2)$

Put the terms of each bracket on the outside of the grid as shown

×	2x	−3
3x	$6x^2$	−9x
+2	4x	−6

Collecting up all the terms inside the grid we have:

$6x^2 - 9x + 4x - 6$

Now simplify, which gives us $6x^2 - 5x - 6$

<u>Another way of expanding brackets with two binomial expressions is by using the FOIL method as shown below:</u>

Example 1: Expand **(x + 3)(x +2)**

Using the **FOIL** method we have:

Step 1: Multiply the **F**irst terms in each bracket to get x×x=x^2

Step 2: Multiply the two **O**utside terms in **(x + 3)(x +2)** =2x

Step 3: Multiply the two **I**nside terms in **(x + 3)(x +2)**= 3x

Step 4: Multiply the two last terms in **(x + 3)(x +2)**= 6

Step 5: Now simply add all the terms and simplify to get:

$x^2 + 2x + 3x + 6 = \mathbf{x^2 + 5x + 6}$

To remember the **FOIL** method just remember the letters of FOIL: **FIRST, OUTSIDE, INSIDE, LAST**

You can see there are several methods, the grid method, the FOIL method or the expansion method as shown below:

(Multiply the first term of the first bracket by the second bracket and then multiply the second term of the first bracket by the second bracket. Finally simplify the expression.)

So $(x + 3)(x + 2) = x(x + 2) + 3(x + 2) = x^2 + 2x + 3x + 6 = x^2 + 5x + 6$

Example 2: Expand $(2x - 1)(x - 2)$

This equals $2x(x - 2) - 1(x - 2) = 2x^2 - 4x - x + 2 = 2x^2 - 5x + 2$

<u>**Typical exam questions**</u>

Example 3: Expand $(y^2 + x)(y - x^2)$

$= y^2(y - x^2) + x(y - x^2) = y^3 - y^2x^2 + xy - x^3$

Example 4: expand and simplify the expression (each bracket has a trinomial expression). However, we can still use the above method:

$(x^2 + 2x + 1)(2x^2 + 3x - 2)$

The principle is the same. Take each term in the first bracket and multiply it out by the second bracket. Finally simplify as much as you can.

So $(x^2 + 2x + 1)(2x^2 + 3x - 2) = x^2(2x^2 + 3x - 2) + 2x(2x^2 + 3x - 2) + 1(2x^2 + 3x - 2) = 2x^4 + 3x^3 - 2x^2 + 4x^3 + 6x^2 - 4x + 2x^2 + 3x - 2$

This simplifies to: $2x^4 + 7x^3 + 6x^2 - x - 2$

Example 5: Expand and simplify the expression $(x + y)^2(2x + 3y)$

$(x + y)^2(2x + 3y) = (x+ y)(x + y)(2x + 3y) = (x^2 + xy + y^2)(2x + 3y)$

$= 2x^3 + 2x^2y + + 2xy^2 + 3x^2y + 3y^2x + 3y^3$

$= 2x^3 + 5x^2y + 5y^2x + 3y^3$

Difference of two squares

Something useful to remember is the <u>difference of two squares:</u>

$p^2 - q^2 = (p + q)(p - q)$

Since $(p + q)(p - q) = p(p - q) + q(p - q) = p^2 + pq - pq - q^2$
$= = p^2 - q^2$

So for example: $16x^2 - 9y^2 = (4x - 3y)(4x + 3y)$

Note:

There are several ways of writing a multiplication:

So you can have either 5×4 = 20 or 5.4 = 20

Similarly, in algebraic expressions a×b is the same as a.b

You can use brackets as well:

(-4)(-5) = 20

(-4)(5) = -20

(-4)(0) = 0 0r (-4).0 = 0

Also the symbol: $\implies$ means 'it implies that' or 'it follows that'

Simplifying Algebraic Fractions

Example 1: Simplify $\dfrac{1}{x+3} + \dfrac{2}{3}$

Method: First find the common denominator which is $3(x+3)$

Then treat it like you were simplifying a fraction

$$\dfrac{1}{x+3} + \dfrac{2}{3} = \dfrac{1\times 3 + 2(x+3)}{3(x+3)} = \dfrac{3+2x+6}{3(x+3)} = \dfrac{9+2x}{3(x+3)}$$

Example 2: Simplify $\dfrac{2}{x-3} - \dfrac{1}{5}$

As before $\dfrac{2}{x-3} - \dfrac{1}{5} = \dfrac{5\times 2 - 1(x-3)}{5(x-3)} = \dfrac{10-x+3}{5(x-3)} = \dfrac{13-x}{5(x-3)}$

Example 3: Simplify $\dfrac{2}{x-2} + \dfrac{3}{x+2}$

Using the method before we get:

$$\dfrac{2}{x-2} + \dfrac{3}{x+2} = \dfrac{2(x+2)+3(x-2)}{(x-2)(x+2)} = \dfrac{2x+4+3x-6}{x^2-4} = \dfrac{5x-2}{x^2-4}$$

<u>Note</u>: $(x+2)(x-2) = x^2 + 2x - 2x - 4 = x^2 - 4$

Example 4: Simplify $\dfrac{x^2+2x-8}{x-2} \div \dfrac{x^2+4x}{x+2}$

Simplifying we get $\dfrac{(x+4)(x-2)}{x-2} \times \dfrac{\cancel{x+2}}{\cancel{x^2+4x}} = \dfrac{(x+4)\cancel{(x-2)}}{\cancel{x-2}} \times \dfrac{x+2}{x(x+4)}$

Cancelling down as shown above we finally get: $\dfrac{x+2}{x}$

Algebra Section 1: Practice Questions

Simplify the following:

(1) $\dfrac{x^3}{y^2} \div \dfrac{y^3}{x}$

(2) $4(3x-7) - 5(2x+3)$

(3) $(2x-3)(2x+3)$

(4) $(x-1)^2(x+1)$

(5) $(x^2 + 3x - 1)(3x^2 + 1)$

(6) $(y^2 - 2y + 3)(3y^2 + 3y - 1)$

(7) $\dfrac{1}{x-1} + \dfrac{3}{7}$

(8) $\dfrac{3}{x-3} - \dfrac{1}{6}$

(9) $\dfrac{3}{x-5} + \dfrac{2}{x+5}$

Answers to Simplifying Expressions

Simplify:

(1) Answer: $\dfrac{x^4}{y^5}$

(2) Answer: $2x - 43$

(3) Answer: $4x^2 - 9$

(4) Answer: $x^3 - x^2 - x + 1$

(5) Answer: $3x^4 + 9x^3 - 2x^2 + 3x - 1$

(6) Answer: $3y^4 - 3y^3 + 2y^2 + 11y - 3$

(7) $\dfrac{4+3x}{7(x-1)}$

(8) $\dfrac{21-x}{6(x-3)}$

(9) $\dfrac{5(x+1)}{x^2-25}$

Algebra Section 2

Factoring

Example 1: Factor the expression: $3x^2 - 6xy$

$= 3x(x - 2y)$ (find the common factor which is 3x in this case)

Example 2: Factor: $3t^2y - 9t^3$

$= 3t^2(y - 3t)$

Example 3: Factor: $x^2 - 2x - 15$

Find two brackets which when multiplied out together gives you the above expression. We find that $x^2 - 2x - 15 = (x - 5)(x + 3)$

Hence the factors are: **(x − 5)(x + 3)**

Note: We will look at factoring cubic expressions later!

Algebra Section 2: Practice Questions

Factor the following algebraic expressions:

(1) $4xp^2 - 3x^2p^3$

(2) $5x^3 - 15x^2 + 25x^4$

(3) $8t^3y^2 - 64t^2y^3 + 16t^2y^2$

(4) $y^2 - 2y - 35$

(5) $6y^2 + 3y - 9$

(6) $7y^2 - 6y - 1$

(7) $9y^2 - 1$

(8) $4n^2 + 12n + 9$

(9) $18p^2 - 3pq - q^2$

Answers to factoring expressions:

(1) $xp^2(4 - 3xp)$

(2) $5x^2(x - 3 + 5x^2)$

(3) $8t^2y^2(t - 8y + 2)$

(4) $(y + 5)(y - 7)$

(5) $(2y + 3)(3y - 3)$

(6) $(7y + 1)(y - 1)$

(7) $(3y + 1)(3y - 1)$

(8) $(2n + 3)(2n + 3)$

(9) $(6p + q)(3p - q)$

Algebra Section 3

Algebraic Substitution and Formula

This is the process of substituting numbers for letters and working out value of the corresponding expression. Some examples that will clarify the process.

Example 1: If k=6 and t=8 work out 2(4k–2t) +kt

Substituting the values of k and t we have:

2(4 × 6–2 × 8) + 6 ×8

=2 × (24 – 16) +48 = 2 ×8 +48 =16+48 =64

So 2(4k – 2t) + kt = 64

Example 2: If t=9 and u= 6 work out $3t^2 - 5u$

Substituting appropriately we get:

$3 \times 9^2 - 5 \times 6$ = 3 × 81–30 =243 – 30 =213

So, $3t^2 - 5u$ =213

(Notice in all these examples we use the **PEMDAS** rule to work out the answers)

PEMDAS: **Parantheses first, then exponents followed by multiplication, division, addition and subtraction**

Easy way to remember **PEMDAS: "Please Excuse My Dear Aunt Sally"**. Which stands for "**Parentheses, Exponents, Multiplication, Division Addition and Subtraction**".

Formula

A formula describes the relationship between two or more variables. You have seen some examples above already. Now let us consider some practical examples.

Example:

The formula for converting the temperature from Celsius to Fahrenheit is given by the formula: $F = \frac{9}{5}C + 32$

(where C is the temperature in degrees Centigrade)

If the temperature is 10 degrees Celsius then what is the equivalent temperature in Fahrenheit?

Using the formula $F = \frac{9}{5}C + 32$, and substituting 10 in place of C, we have $F = \frac{9}{5} \times 10 + 32 = \frac{90}{5} + 32 = 18 + 32 = 50$. Hence, 10 degrees centigrade = 50 degrees Fahrenheit

Explanation of working out above: Remember we multiply and divide before adding and subtracting) There are no parantheses to worry about. When working out $\frac{9}{5} \times 10 + 32$, multiply 9 by 10 to get 90, divide this by 5 to get 18, finally add 18 and 32 together to get 50

Example 3: Convert 68 degrees Fahrenheit to degrees Celsius. The formula for converting the temperature from Fahrenheit to Celsius is given by:

$C = \frac{5}{9}(F-32)$, So to change 68 degrees Fahrenheit to degrees Celsius we can substitute for F in the formula $C = \frac{5}{9}(F-32)$, $C = \frac{5}{9}(68-32) =$

$\frac{5}{9} \times 36 = 5 \times 4 = 20$. Hence, 68 degrees Fahrenheit = 20 degrees Celsius

Explanation of the working out above: Using PEMDAS we work out the parentheses first. This gives us 68-32 =36. We now divide this by 9 and multiply by 5. Clearly 36÷9 =4 and finally 5×4 =20

Changing the subject of a formula

(Changing the subject of a formula is basically re-arranging formulas)

Example 1: In the formula $a = bx + c$ make x the subject of the formula

Method: Apply the same rules as you would to equations

In this case subtract c from both sides to get $a - c = bx$

Now divide both sides by b to get $\frac{a-c}{b} = x$

In other words $x = \frac{a-c}{b}$

Example 2: In the formula $\frac{ay^2}{b} + m = k$, make y the subject of the formula

Method:

Step 1: Subtract m from both sides to get $\frac{ay^2}{b} = k - m$

Step 2: Multiply both sides by 'b' to get $\frac{ay^2}{\cancel{b}} \times \cancel{b} = b(k - m)$

(The 'b's on the left hand side cancel)

Hence we now have: $ay^2 = b(k - m)$

Step 3: divide both sides by 'a' (to cancel the 'a' on the left hand side)

We now have $y^2 = \frac{b(k-m)}{a}$

Step 4: Take the square root of both sides to get $y = \sqrt{\frac{b(k-m)}{a}}$

So we finally find that $y = \sqrt{\frac{b(k-m)}{a}}$

Example 3: In the formula $\frac{a}{1+t^2} = b + c$, make t the subject of the formula

Step 1: Multiply both sides by $1 + t^2$ to get $a = (b + c)(1 + t^2)$

Step 2: Divide both sides by (b + c) to get: $\frac{a}{b+c} = 1 + t^2$

Step 3: Subtract '1' from both sides to get: $\frac{a}{b+c} - 1 = t^2$

Step 4: This simplifies to $\frac{a-1(b+c)}{b+c} = t^2$, which simplifies to $\frac{a-b-c}{b+c} = t^2$

Step 5: Take the square root of both sides: $\sqrt{\frac{a-b-c}{b+c}} = t$

So finally we have $t = \sqrt{\frac{a-b-c}{b+c}}$

Algebra Section 3: Practice Questions

Practice Questions on Change the Subject of formula

(1) $C = \frac{5}{9}(F - 32)$, make F the subject of the formula

(2) $x = \frac{m(s-t)}{x}$, make x the subject of the formula

(3) $y^3 = \frac{3(y^3 + t)}{m}$, make y the subject of the formula

(4) In the formula $\frac{1}{f} = \frac{1}{u} + \frac{1}{v}$ make v the subject of the formula

(5) Given that $s = ut + \frac{1}{2}at^2$, make a the subject of the formula

(6) If $\frac{c}{1+x} = m + n$, make x the subject of the formula

Answers to change the subject of formula:

(1) $F = \dfrac{9C}{5} + 32$

(2) $x = +/- \sqrt{m(s-t)}$

(3) $y = \left(\dfrac{3t}{m-3}\right)^{\frac{1}{3}}$

(4) $v = \dfrac{fu}{u-f}$

(5) $a = \dfrac{2(s-ut)}{t^2}$

(6) $x = \dfrac{c-m-n}{m+n}$

Algebra Section 4

Solving equations

(An unknown such as 'x' is called a variable)

Example 1: Solve the equation $5x - 1 = 2x + 8$

First add 1 to both sides, which gives:

$5x = 2x + 9$

Now subtract 2x from both sides to give $3x = 9$

Finally divide both sides by 3 to get $x=3$.

(Notice each step simplifies the equation further)

Example 2: Solve the equation $5(2x + 1) = 4(2x + 1)$

To solve this first multiply out the bracket which gives:

$10x + 5 = 8x + 4$

(Multiply each term outside the bracket by each term inside the bracket)

Now subtract 5 from both sides, which gives:

$10x = 8x - 1$

Now subtract 8x from both sides, which gives:

$2x = -1$

Finally, divide both sides by 2 to get $x = -1/2$ or -0.5

Example 3: Solve the equation $\frac{2x}{3} + 5 = 7$

We can simplify this to $\frac{2x}{3} = 2$ (by subtracting 5 from both sides)

Now multiply both sides by 3 to get the expression below:

$2x = 6$, so $x = 3$

Example 4

Solve the equation $\sqrt{4 - \frac{x+3}{3x+2}} = 3$

Although this might look complicated the basic rule is whatever you do to one side you must do the same to the other.

Step 1: Square both sides so we get $4 - \frac{x+3}{3x+2} = 9$

Step 2: Cross –multiply everything by the denominator (3x + 2)

We get: $4(3x + 2) - (x + 3) = 9(3x + 2)$

Simplify to get $12x + 8 - x - 3 = 27x + 18$

Simplify further to get $11x + 5 = 27x + 18$

Subtract 18 from both sides to get $11x - 13 = 27x$

Now subtract 11x from both sides to get $-13 = 16x$

Which is the same as 16x = -13, this means $x = \frac{-13}{16}$

Solving linear equations with inequalities

Example 1: Solve the inequality 2x +5>9

This simply says 2x + 5 is greater than 9. To find x still use the rules of a simple equation. That is, whatever you do to one side you must do to the other.

If 2x +5>9, then 2x >4 (by taking away 5 from both sides)

Now, divide both sides by 2 to get x >2. Our answer for x is all values greater than 2.

Example 2: Solve the inequality 2(5x – 1) ≥ 3x + 14

Method:

This simplifies to 10x – 2 ≥ 3x + 14

Subtract 3x from both sides to get 7x – 2 ≥ 14

Add 2 to both sides to get 7x ≥ 16

Dividing both sides by 7 to get $x \geq \frac{16}{7}$ $\implies$ $x \geq 2\frac{2}{7}$

Example 3: Solve the inequality 4 – 2x < 16

$\implies$ -2x < 12 $\implies$ -x < 6 $\implies$ x > -6

(Note: In inequalities when you divide both sides by -1 you also change the sign of the inequality)

Example 3: Show the inequality -2 < x ≤ 2 by way of a number line.

The answer is shown below:

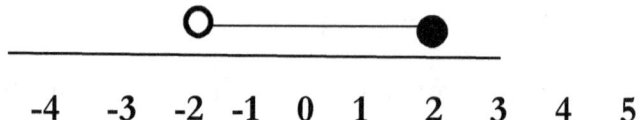

```
 -4  -3  -2  -1  0  1  2  3  4  5
```

(Note the convention that a dark shaded circle implies 2 is included because x ≤ 2 and at -2 unshaded (open) circle implies -2 is not included because it is <)

Word Problems using Algebra

Examples:

(1) Fatima and Louise have $350 between them. Louise has $80 less than Fatima. How much do they each have?

Method: Let the amount Fatima has be represented by x
Hence, Louise has x – 80. We know that the sum of the two amounts = $350. That is x + x – 80 = 350. Simplifying, we get 2x – 80 = 350. Now add 80 to both sides so we have 2x - 80 +80 = 350 + 80. Which means 2x = 430, or x = 215. This means Fatima has $215 and Louise has $135 (Since Louise has $80 less than Fatima)

(2) The cost of a coat after a 20% discount is $85. What was its original price?

Method: Let the original price be $x. This means x − 20% of x = 85. Or x − 0.2x = 85, which simplifies to 0.8x = 85. Now divide both sides by 0.8. So we get x = 85÷0.8 = 106.25. Hence the original price is $106.25

(3) The area of a rectangle is $162 m^2$. The length of the rectangle is two times the width. What is the length and width of the rectangle?

Method: Let the width =w, hence the length = 2w. We know that the area of a rectangle is length × width = 2w×w = $2w^2$. The area of the rectangle is given as $162 m^2$. Hence, $2w^2 = 162$. Dividing both sides by 2, we get $w^2 = 81$. Hence w = $\sqrt{81}$ =9. So the width is 9m and the length is 18m.

(4) John's annual salary is $\frac{3}{4}$ of Hilary's salary. Hilary's salary is twice Betty's. The total salary between them is $450,000. How much did each of them earn?

Method: Let Hilary's salary be x (in dollars). Hence, John's salary is $\frac{3}{4}$ x. Also, since Hilary earns twice as much as Betty, then Betty earns half of Hillary's = $\frac{1}{2}$x. Finally, we know that x + $\frac{3}{4}$ x + $\frac{1}{2}$x = $450,000, simplifying $2\frac{1}{4}$x = 450,000. Or, $\frac{9}{4}$ x = 450,000. This means 9x = 1,800,000 or x = 200,000. So Hilary earns $200,000, John earns $150,000 (three quarters of Hilary's amount) and Betty earns $100, 000 (half of Hilary's salary)

(5) The sum of two numbers is 30 and the difference between them is 8. What are the two numbers?

Method: Let the unknown numbers be x and y. This means x + y = 30, and x − y = 8, If we add the above two equations we get 2x = 38 (The y's cancel). Hence x = 19 and y = 11

(6) The second number is $\frac{3}{4}$ of the first number. The sum of two numbers is 5.25. What are the two numbers?

Method: The sum of the two numbers is 5.25. Let one of the numbers be x. Hence, x + $\frac{3}{4}$x = 5.25. Simplifying, we get 1.75x = 5.25, dividing both sides by 1.75 we get x = 3. So, $\frac{3}{4}$x = 0.75 × 3 = 2.25. So the two numbers are 3 and 2.25.

Algebra Section 5

Algebraic Proofs

Example 1: Prove that $(3n + 1)^2 - (3n - 1)^2$ is a multiple of 4.

Proof: $(3n + 1)^2 = 9n^2 + 6n + 1$ and $(3n - 1)^2 = 9n^2 - 6n + 1$

So $(3n + 1)^2 - (3n - 1)^2 = 9n^2 + 6n + 1 - (9n^2 - 6n + 1) =$

$9n^2 + 6n + 1 - 9n^2 + 6n - 1 = 12n = 4(3n)$ which is a multiple of 4

Example 2: Prove that when n and m are positive integers that 2n + 1 + 2m + 1 is always even.

Proof: 2n + 1 + 2m + 1 = 2n + 2m + 2 = 2(n+m) + 2

Let n + m = k so we have 2(n+m) + 2 = 2k + 2 = 2(k+1) which is even

Example 3: Prove that $(n + 1)^2 - (n - 5)^2 = 12(n - 2)$

Proof: $(n + 1)^2 - (n - 5)^2 = n^2 + 2n + 1 - (n^2 - 10n + 25) =$

$n^2 + 2n + 1 - n^2 + 10n - 25 = 12n - 24 = 12(n - 2)$

<u>Remember for all integer values of n:</u>

<u> 2n is even and 2n + 1 is odd</u>

Algebra Section 5: Practice Questions

Practice Questions on Algebraic Proofs:

(1) Prove that $\frac{1}{4}(2n+1)(n+4) - \frac{1}{4}n(2n+1) = 2n + 1$

(2) Prove that $(n+2)^2 - (n-6)^2 = 16(n-2)$

(3) Prove that $(n+6)^2 - n(n-3)$ is a multiple of 3 for all positive integers n.

(4) Prove that $(n+1)^2 - (n-5)^2 = 12(n-2)$

(5) Prove that $(3n+1)^2 - (3n-1)^2$ is a multiple of 12 for all positive integers n

(6) Prove that $(p+1)^2 - (p-1)^2 + 1$ is always odd for positive integers

(7) Prove that the difference between the squares of any two consecutive even numbers is always a multiple of 4

Answers to Algebraic Proofs:

(1) $\frac{1}{4}(2n+1)(n+4) - \frac{1}{4}n(2n+1)$

$= \frac{1}{4}(2n^2 + 9n + 4) - \frac{1}{4}(2n^2 + n)$

$= \frac{1}{4}(2n^2 + 9n + 4 - 2n^2 - n) = \frac{1}{4}(8n + 4) = \frac{4}{4}(2n + 1) = 2n + 1$

Hence $\frac{1}{4}(2n+1)(n+4) - \frac{1}{4}n(2n+1) = 2n + 1$

Better method: $\frac{1}{4}(2n+1)(n+4) - \frac{1}{4}n(2n+1) = \frac{1}{4}(2n+1)(n+4-n) = \frac{1}{4}(2n+1) \times 4 = 2n + 1$

Note: $\frac{1}{4}(2n+1)$ is the common factor.

(2) $(n+2)^2 - (n-6)^2 = n^2 + 4n + 4 - n^2 + 12n - 36$
$= 16n - 32 = 16(n-2)$

Hence $(n+2)^2 - (n-6)^2 = 16(n-2)$

(3) $(n+6)^2 - n(n-3) = n^2 + 12n + 36 - n^2 + 3n$
$= 15n + 36 = 3(5n + 12)$
Hence, $3(5n+12)$ is a multiple of 3 for all positive integers of n

(4) $(n+1)^2 - (n-5)^2 = n^2 + 2n + 1 - (n^2 - 10n + 25)$
$= n^2 + 2n + 1 - n^2 + 10n - 25 = 12n - 24 = 12(n-2)$

Hence $(n+1)^2 - (n-5)^2 = 12(n-2)$

(5) $(3n+1)^2 - (3n-1)^2 = 9n^2 + 6n + 1 - (9n^2 - 6n + 1) = 9n^2 + 6n + 1 - 9n^2 + 6n - 1 = 12n$ (which is a multiple of 12)

(6) $(p+1)^2 - (p-1)^2 + 1 = p^2 + 2p + 1 - (p^2 - 2p + 1) + 1 = p^2 + 2p + 1 - p^2 + 2p - 1 + 1 = 4p + 1$ which is odd

(7) Let 2p and 2p + 2 be two consecutive even numbers
Hence $(2p+2)^2 - (2p)^2 = 4p^2 + 8p + 4 - 4p^2 = 8p + 4 = 4(2p+1)$ which is a multiple of 4

Algebra Section 6

Simultaneous Equations

We saw earlier that simple equations allow us to solve problems involving one unknown. When you have to solve problems involving more than one unknown you need more than one equation to solve these.

A simultaneous equation with two variables (meaning two unknowns), say x and y, typically involves two equations. The problem is then to find the values of x and y which satisfies the equations at the same time. Another way of saying simultaneous is 'at the same time'.

We will first consider one traditional method of solving simultaneous equations.

Example 1: Solve the simultaneous pair: $2x + 4y = 5$

$3x + 5y = 9$

First let us understand the problem. The problem is to find the values of x and y such that the equations are true.

Method 1:

Consider the substitution method:

We will express x in term of y in the first equation and substitute for x in the second equation. We have:

$2x + 4y = 5$

$3x + 5y = 9$

From the first equation we have 2x =5 – 4y (subtract 4y from both sides)

So x =2.5 –2y (we get this by dividing both sides of the previous expression by 2.)

Now substitute x =2.5 –2y in the second equation, which gives us:

3(2.5 – 2y) + 5y =9

So, 7.5 –6y +5y =9

So, 7.5 – y =9

So, 7.5 – 9= y

Which gives y = –1.5

Now we need to find x. We can substitute the value of y in equation 1 to find x.

The first equation is 2x + 4y =5

Substituting for y, we get 2x + 4×(– 1.5) =5

which means 2x – 6 =5, So 2x = 11, hence x = 5.5

Check

We can check in equation 1 to see if the values we found satisfy the equation.

The first equation is: 2x +4y =5

Substituting, x=5.5 and y = -1.5 we get:

2×5.5 + 4×(-1.5) = 11 – 6 = 5 as required.

Method 2: Eliminate one of the variables

Consider the simultaneous pair of equations as before:

2x + 4y = 5   (1)

3x + 5y = 9   (2)

Try and make the x or the y terms the same and then add or subtract the equations to eliminate one of the variables

Suppose we make the 'x' term the same. We multiply equation (1) by 3 and equation (2) by 2. The new equations are now shown below:

6x + 12y = 15   (3)

6x + 10y = 18   (4)

Now subtract (4) from (3) and we get:

6x – 6x + 12y – 10y = 15 – 18

2y = -3 hence y = -3/2 = –1.5

Now substitute y = -1.5 in equation (1) and we get:

2x + 4×(-1.5) = 5 which means 2x – 6 = 5 or 2x = 11 so x = 11/2 or 5.5

Hence as before x = 5.5 and y = –1.5

There is another method which involves drawing graphs of the two equations and finding the point at which they intersect.

Simultaneous equations with three unknowns:

Solve the simultaneous equations:

$2x + 3y + z = 4$ ....(1)

$3x - 3y + 2z = 2$ .... (2)

$2x + 3y - z = 2$ ....(3)

If we add (1) & (2) and adding (2) & (3) we can eliminate y.

So we now have $5x + 3z = 6$ ....(4)

$\quad\quad 5x + z = 4$............(5)

We can now subtract (5) from (4) to get $2z = 2 \quad\quad z = 1$

Substitute $z = 1$ in (5) $\implies 5x + 1 = 4 \implies x = \dfrac{3}{5}$

Finally substitute $z = 1$ and $x = \dfrac{3}{5}$ in (1) to get $\dfrac{6}{5} + 3y + 1 = 4$

$\implies 3y = 3 - \dfrac{6}{5} \implies 3y = \dfrac{15-6}{5} \implies 3y = \dfrac{9}{5} \implies y = \dfrac{3}{5}$

Hence the solution to the simultaneous equations with three unknowns in this case are $x = \dfrac{3}{5}$, $y = \dfrac{3}{5}$ and $z = 1$

Practice Questions Algebra Sections 5 and 6

(1) Solve the following equations

(a) $3(2x + 4) = 2(5x - 7)$

(b) $2x + 7 \geq 2$

(c) $2(5x - 7) > 6(2x - 8)$

(d) $\dfrac{1}{3-x} = \dfrac{3}{x+5}$

(2) Solve the following simultaneous equations

(a) $3x - 2y = 5$,

$2x + 3y = 7$

(b) $2x - y = 2$

$5x + y = 8$

(c) $x + 2y + z = 1$

$x + 3y - z = 3$

$2x - 2y + z = 7$

(3) Solve these word problems algebraically

(a) John and Elizabeth have $375 between them. John has $85 less than Elizabeth. How much do they each have?

(b) The sum of two numbers is 35 and the difference between them is 6. What are the two numbers?

(4) Solve the inequality $2(11x - 3) \geq 25x + 17$

(5) Show the inequality $-1 < x \leq 2$ by way of a number line.

(6) Solve the equation $\sqrt{2 - \frac{2x+2}{4x+1}} = \sqrt{6}$

Answers to Algebra Sections 5 and 6

1. (a) $x = 6.5$

   (b) $x \geq -\dfrac{5}{2}$ or $x \geq -2.5$

   (c) $x < 17$

   (d) $x = 1$

2. (a) $x = \dfrac{29}{13}, y = \dfrac{11}{13}$

   (b) $x = = \dfrac{10}{7}, y = = \dfrac{6}{7}$

   (c) $x = \dfrac{46}{13}, y = -\dfrac{8}{13}, z = -\dfrac{17}{13}$

3. (a) John has $145 and Elizabeth $230

   (b) The two numbers are 14.5 and 20.5

4. $x \leq -\dfrac{23}{3}$ or $-7\dfrac{2}{3}$

5. The answer is shown below:

   -4 -3 -2 -1 0 1 2 3 4 5

6. $x = 1$

Algebra Section 7

Solving Quadratic Equations

For the general quadratic equation $ax^2 + bx + c = 0$

The formula for solving the equation is given by: $\mathbf{x} = \dfrac{-b \pm \sqrt{b^2 - 4ac}}{2a}$

We can show that this is true by the method of completing the square as shown below.

Consider the general quadratic equation $ax^2 + bx + c = 0$

Dividing through by 'a' we get:

$$x^2 + \frac{b}{a}x + \frac{c}{a} = 0$$

Now we use the method of completing the square

First halve the middle term coefficient and then square the expression on the left hand side as shown below:

$$\left(x + \frac{b}{2a}\right)^2 = x^2 + \frac{b}{a}x + \frac{b^2}{4a^2}$$

Adjusting to get the original expression, we have:

$$\left(x + \frac{b}{2a}\right)^2 - \frac{b^2}{4a^2} + \frac{c}{a} = x^2 + \frac{b}{a}x + \frac{c}{a}$$

We can write $\left(x + \dfrac{b}{2a}\right)^2 - \dfrac{b^2}{4a^2} + \dfrac{c}{a} = 0$

$$\left(x + \frac{b}{2a}\right)^2 = \frac{b^2}{4a^2} - \frac{c}{a}$$

Simplifying the right hand side we get:

$$\left(x + \frac{b}{2a}\right)^2 = \frac{b^2}{4a^2} - \frac{4ac}{4a^2}$$

$$\left(x + \frac{b}{2a}\right)^2 = \frac{b^2 - 4ac}{4a^2}$$

$$x + \frac{b}{2a} = \pm \frac{\sqrt{b^2 - 4ac}}{2a}$$

$$x = -\frac{b}{2a} \pm \frac{\sqrt{b^2 - 4ac}}{2a}$$

$$x = \frac{-b \pm \sqrt{b^2 - 4ac}}{2a}$$

Example:

Solve the equation $2x^2 - 5x + 2 = 0$ using the quadratic formula.

Method:

When the above equation is compared to the general equation $ax^2 + bx + c = 0$

We can see that a = 2, b= -5 and c =2

Since, $x = \dfrac{-b \pm \sqrt{b^2 - 4ac}}{2a}$

By substituting the above values we can see that:

$$x = \dfrac{-(-5) \pm \sqrt{(-5)^2 - 4 \times 2 \times 2}}{2 \times 2}$$

$$x = \dfrac{5 \pm \sqrt{25 - 16}}{4}$$

$$x = \dfrac{5 \pm \sqrt{9}}{4}$$

$$x = \dfrac{5 \pm 3}{4} = \dfrac{8}{4} \text{ or } \dfrac{2}{4}$$

Hence $x = 2$ or $\dfrac{1}{2}$

(Note: The formula method is particularly useful if you find it hard to factorise or if a quadratic expression cannot be factorised)

Solving quadratic equations using factorisation when possible:

Example 1: Solve the equation $x^2 + 5x + 6 = 0$

We can factorise the above quadratic equation as $(x + 3)(x + 2) = 0$

This means either $x + 3 = 0 \implies x = -3$ or $x + 2 = 0 \implies x = -2$

Example 2: Solve the quadratic equation $2x^2 - 5x + 2 = 0$

We can write the above equation as $(2x - 1)(x - 2) = 0$

(You can do this by trial and error with a little bit of common sense)

For example the only way to get $2x^2$ is by having x and 2x in the two brackets. Also the only way to get $+ 2$ as the last term is to have +1 and +2 or -1 and -2. Finally, as the middle term is -5x the factors have to be $(2x - 1)(x - 2)$

So if $(2x - 1)(x - 2) = 0$ this means either $2x - 1 = 0$ or $x - 2 = 0$

If $2x - 1 = 0 \implies 2x = 1$ and $x = \frac{1}{2}$ and if $x - 2 = 0 \implies x = 2$

Hence the solution to the quadratic equation $2x^2 - 5x + 2 = 0$ is

Either $x = \frac{1}{2}$ or $x = 2$

Things to note in quadratic equations and the quadratic formula:

(1) $ax^2 + bx + c = 0$ is a quadratic equation providing 'a' is not zero.
(2) There are two solutions (or roots) to a quadratic equation
(3) The roots are real so long as in the formula $x = \frac{-b \pm \sqrt{b^2 - 4ac}}{2a}$ the bit inside the square root is > 0. The bit inside the square root, that is $b^2 - 4ac$, is called the

discriminant. Note if $b^2 - 4ac = 0$, there is only one real root

(4) When $b^2 - 4ac < 0$, then the roots are not real.

Below are examples of equations with two solutions (two roots), one solution (one root) and no solution (no real roots)

(a) The equation $x^2 + 2x - 15 = 0$, has two real roots, x= -5 and x = 3 as shown in the graph below

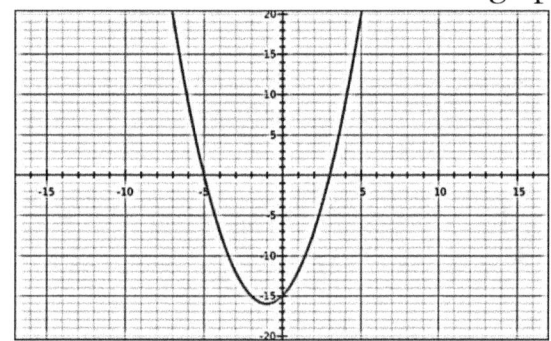

(b) The equation $x^2 - 6x + 9 = 0$ has one real root at x = 3 as shown below

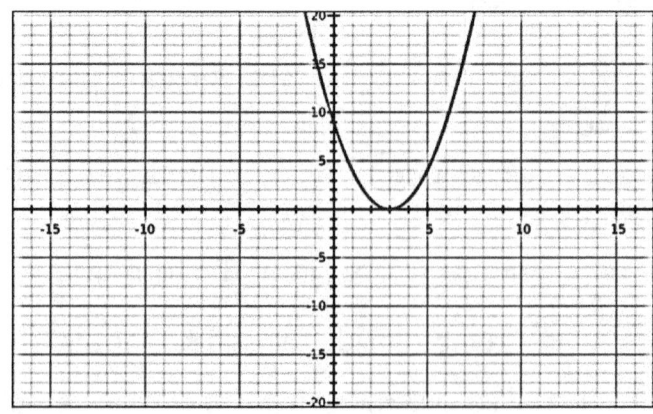

(c) The equation $x^2 - 6x + 12 = 0$ has no real roots as the parabola does not intersect the x-axis at any point.

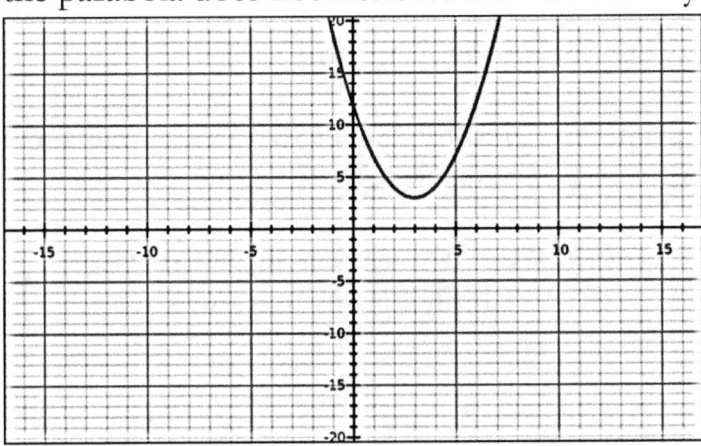

Solving Quadratic Inequalities

Example 1: Solve the quadratic inequality $-x^2 + x + 8 \geq 2$

First re-write this as $-x^2 + x + 6 \geq 0$

Plot its graph and find the values that satisfy this inequality. Namely, values of x, when $y \geq 0$.

First let us see if we can simplify and factorise the equation $-x^2 + x + 6 = 0$. Multiplying through by -1 we get $x^2 - x - 6 = 0$.

$\Rightarrow$ $(x - 3)(x + 2) = 0$ $\Rightarrow$ **x = 3 or -2**. (Also note that in the initial equation the coefficient of x^2 was negative. This implies that the graph is inverted 'U' shaped as shown below)

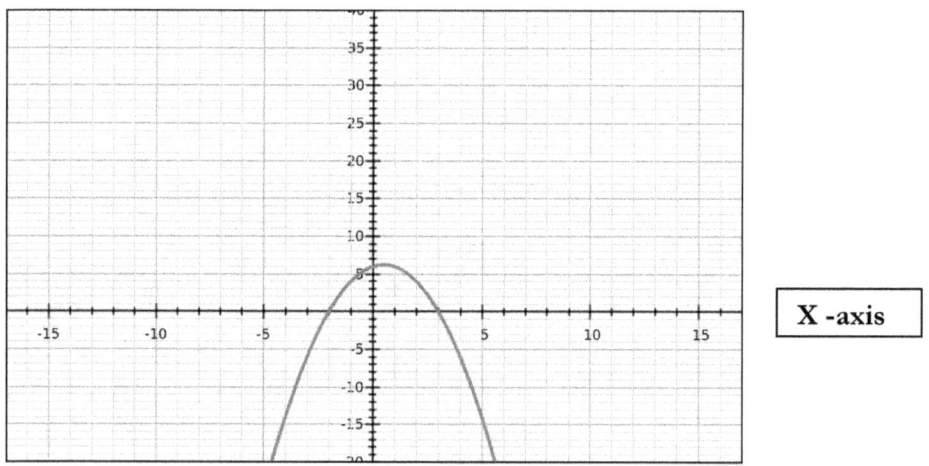

The solution to $-x^2 + x + 8 \geq 2$ is when $-2 \leq x \leq 3$

Example 2:

Solve the inequality $2x^2 + x - 1 > 0$

When it crosses the x-axis we can find the values of x.

That is, $2x^2 + x - 1 = 0$ ➡ $(2x - 1)(x + 1) = 0$

➡ $x = \frac{1}{2}$ or $x = -1$. Also since the coefficient of x^2 is positive then the parabola will be **U-shaped as shown below**.

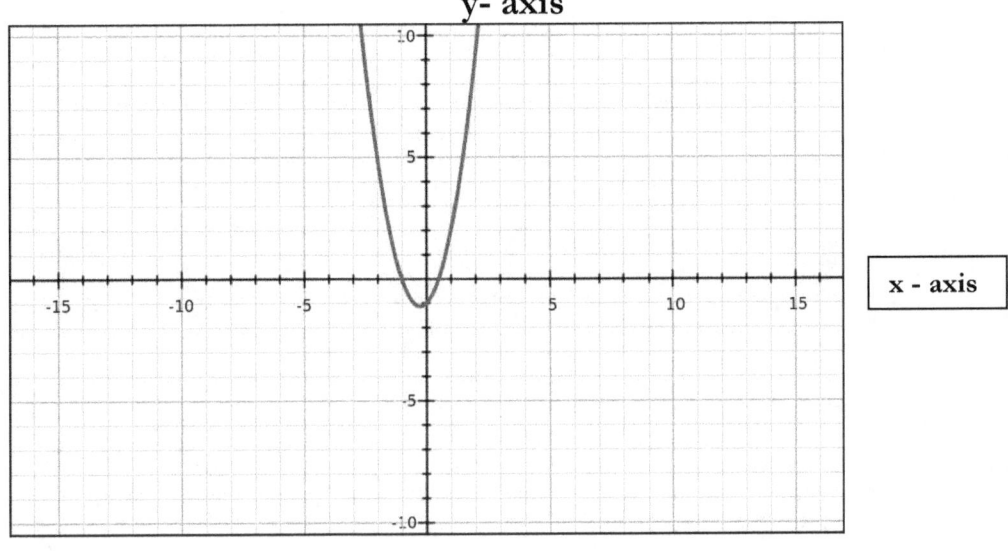

Examining the graph we can see that $2x^2 + x - 1 > 0$ is true when x is less than -1 and when x is greater than $\frac{1}{2}$

That is when $x < -1$ or $x > \frac{1}{2}$

Note: Similar principles have to be applied if you have to solve an quadratic inequality which is $<$ or $\leq$ rather than $>$ or $\geq$.

You can best see this visually by actually drawing the graph(s).

Graphs of quadratic equations

Quadratic Equations

These are of the form $f(x) = ax^2 + bx + c$. Note $f(x)$ is called a function of x since y is defined in terms of x.

Consider the example below: Plot the equation $y = 3x^2 - 2x + 1$

First choose suitable values of x say from -3 to + 3 and find the corresponding values of y as shown in the table below:

x	-3	-2	-1	0	1	2	3
$3x^2$	27	12	3	0	3	12	27
-2x	6	4	2	0	-2	-4	-6
+1	1	1	1	1	1	1	1
Y	34	17	6	1	2	9	22

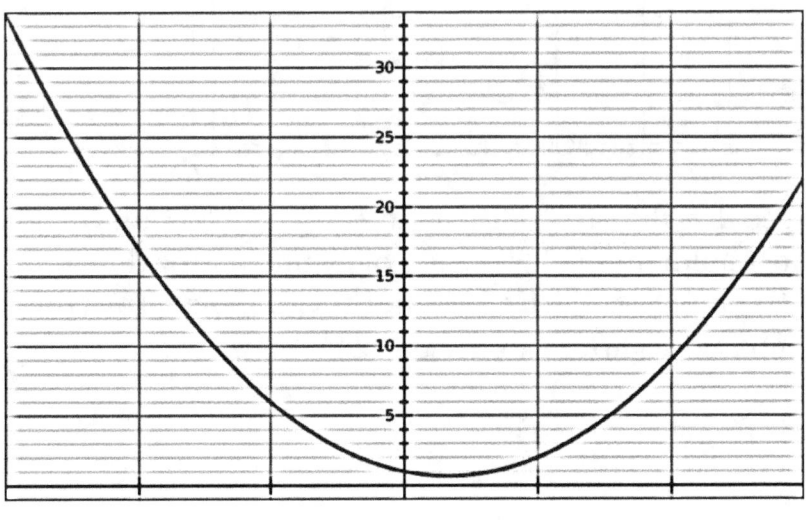

Finding the maximum and minimum values of a parabola (that is the turning points of the graph.

For $f(x) = (x + a)^2 + b$ the turning points are at $x = -a$ and $y = b$

In this case the turning point is a minimum

Example: If $f(x) = (x + 2)^2 + 3$, what are the co-ordinates of the turning points and state whether it is a maximum or a minimum. Also state the line of symmetry.

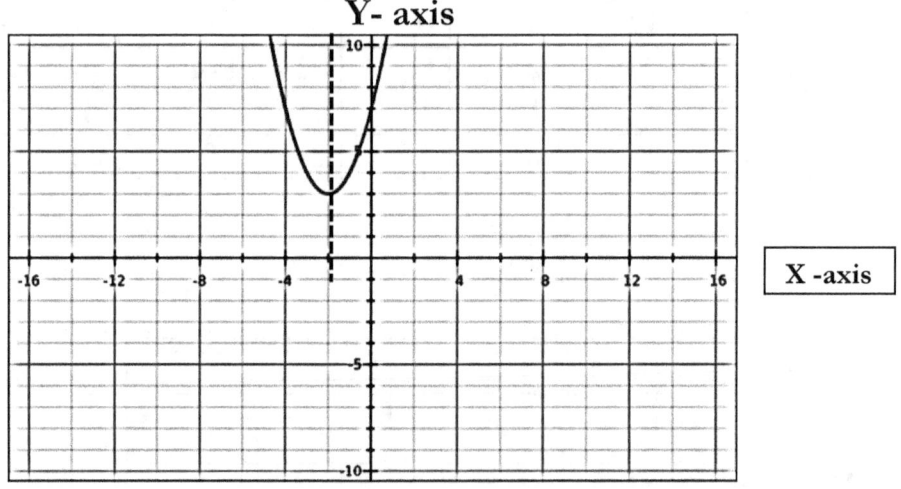

You can see that the graph of $y = (x + 2)^2 + 3$ has a turning point at its minimum -2, 3. The line of symmetry for this function f(x) is at x = -2. (**Remember f(x) means a function of x, which is the same as $y = (x + 2)^2 + 3$**).

Likewise if $f(x) = -(x + a)^2 + b$

Then the turning point is at x = -a, y = b and this time is a maximum.

Example 2:

For the equation $y = -(x + 3)^2 + 2$ find the co-ordinates of the turning point and the line of symmetry.

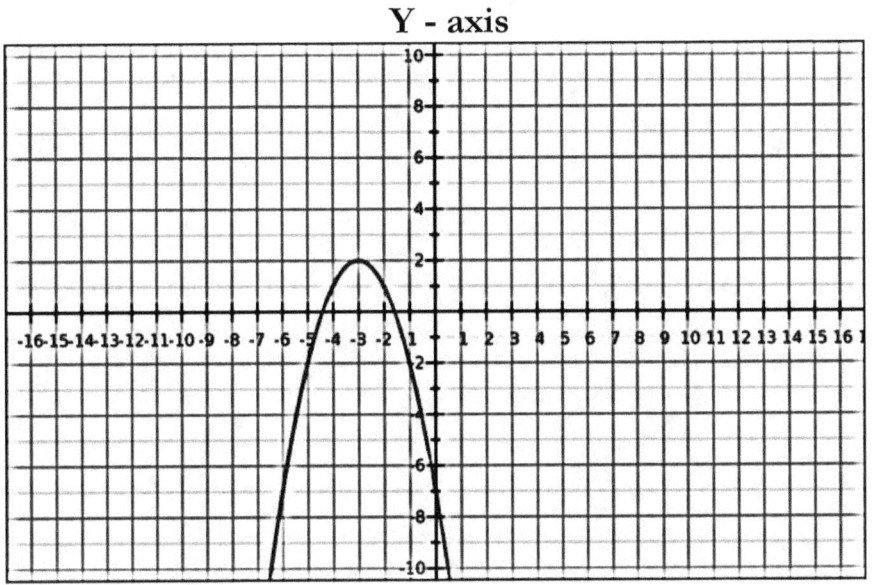

You can see from the graph of $y = -(x + 3)^2 + 2$ above that the turning point is at its maximum where the co-ordinates correspond to (-3, 2). The line of symmetry is at $x = -3$

<u>Note</u>: This means if you are given an equation of a parabola such as $y = x^2 - 2x + 1$, you can re-state it in the form $y = (x + a)^2 + b$, you can then find the co-ordinates of the turning point. In this case the turning point has a minimum value.

Example: Consider the parabola $y = x^2 + 2x + 1$. We can express this as $(x + a)^2 + b = x^2 + 2xa + a^2 + b$. That is the equation of the parabola can be written as $(x + 1)^2 + 0$. Hence the turning point occurs at the point (-1, 0) and it is a minimum.

Cubic equation

Example: Plot the equation $y = x^3 - 1$

x	-3	-2	-1	0	1	2	3
x^3	-27	-8	-1	0	1	8	27
-1	-1	-1	-1	-1	-1	-1	-1
Y	-28	-9	-2	-1	0	7	26

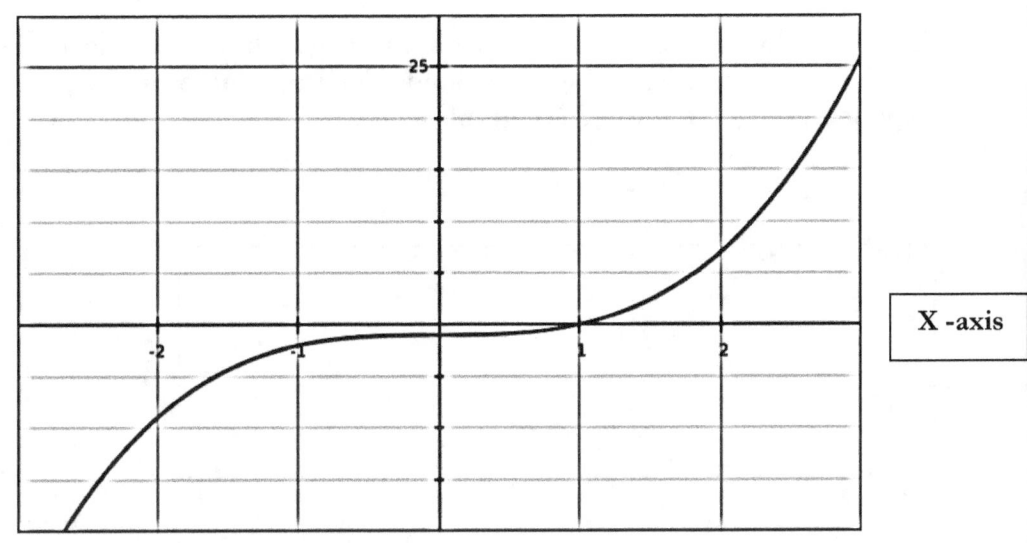

Exponential Graphs

Are of the form y = k to the power of x

Example: To plot the graph of $y = 3^x$ first find some coordinates:

X	-2	-1	0	1	2
y	$3^{-2}=$ 1/9= 0.111	3^{-1} =1/3 =0.333	3^{-0} =$1/3^0$ = $\frac{1}{1} = 1$	$3^1 = 3$	$3^2 = 9$

As we plot these points, you can see that the graph of $y = 3^x$ never falls below the x axis, and when x is positive the y values increase exponentially (or rapidly). See graph below.

Y – axis

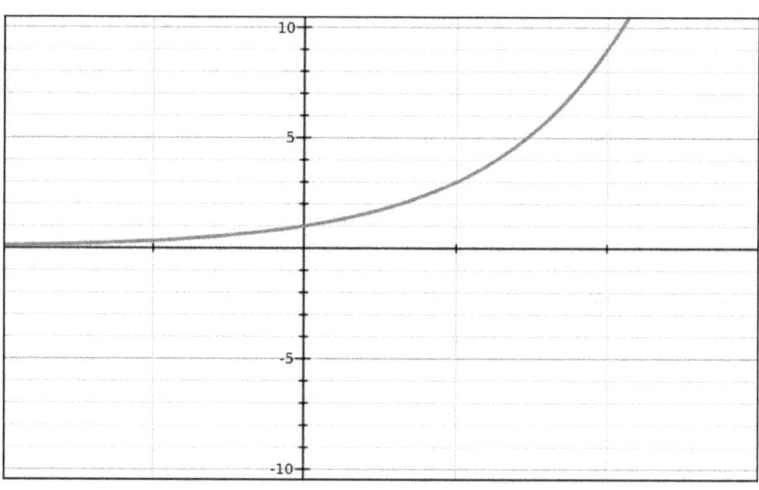

X =axis

Solving equations using graphical methods

Example

You are given that the quadratic equation $y = x^2 - 4x + 8$ and the linear equation $y = 3x - 2$ intersect at two points A and B. Find the co-ordinates of these two points A and B where the equations intersect.

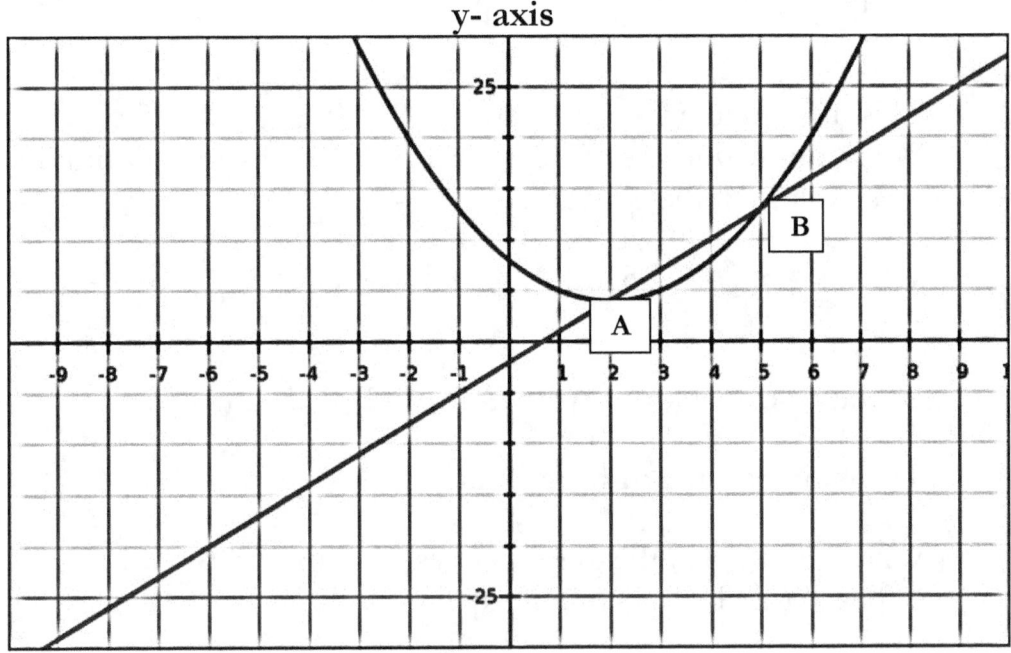

Method: First plot the equations $y = x^2 - 4x + 8$ and the linear equation $y = 3x - 2$ as shown above.

You can see that the co-ordinates of A are (2, 4) and the co-ordinates of B are (5, 13)

Solving equations mathematically when one is linear and the other is quadratic:

The two equations are $y = x^2 - 4x + 8$ and $y = 3x - 2$.

This means $x^2 - 4x + 8 = 3x - 2$

Simplifying this we get $x^2 - 7x + 10 = 0$ ⟹ $(x-5)(x-2) = 0$

This means either, $x - 5 = 0$ or $x - 2 = 0$ ⟹ $x = 5$ or 2. We can now find the corresponding values of y by substituting these values in the equation $y = 3x - 2$

When $x = 5$, $y = 3 \times 5 - 2 = 13$ and when $x = 2$, $y = 3 \times 2 - 2 = 4$

So the co-ordinates of $A = (2, 4)$ and $B = (5, 13)$

Transformations of functions

When $y = f(x + a)$ the graph moves 'a' units left. (It is the opposite of what you might expect)

Likewise when $y = f(x - a)$ the graph moves 'a' units to the right

Consider the two graphs below (1) f(x) and (2) f(x) = f(x +2)

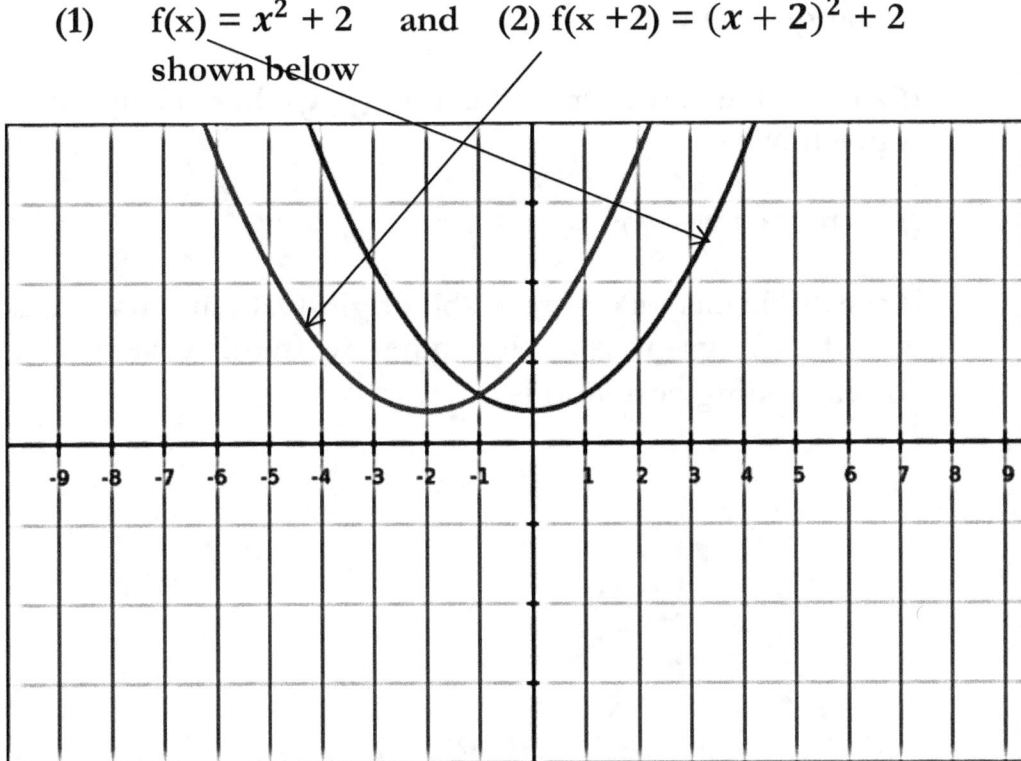

(1) f(x) = x^2 + 2   and   (2) f(x +2) = $(x + 2)^2$ + 2
shown below

You can see that f(x + 2) has shifted to the left by 2 units (this seems counter-intuitive) but f(x + 2) does shift to the left by 2 units and <u>not</u> to the right.

Other types of transformations involving y = f(x)

It is worth remembering that y = f(x) + a moves up the y-axis by 'a' units and likewise y = f(x) – a moves down the y –axis by 'a' units.

Finally y = k×f(x) or kf(x) simply means the graph of f(x) stretches along the y –axis by a factor of k.

Algebra Section 8

Equation of a Circle

If a circle has radius 'r' and centre (p, q) then its equation is given by:

$$(x - p)^2 + (y - q)^2 = r^2$$

If the circle has its centre at the origin (0, 0) and its radius is '1' then its graph is as shown below. In this case the corresponding equation is $x^2 + y^2 = 1$

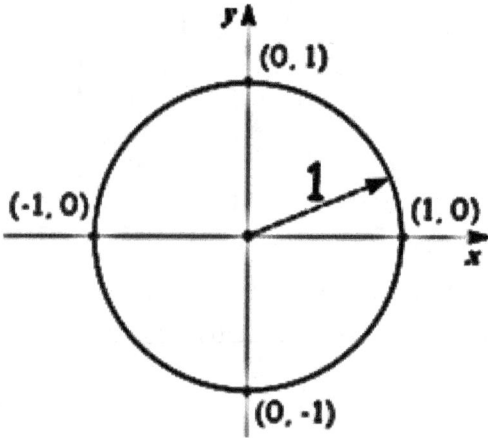

Example 1:

Find the equation of a circle whose radius is 5 and its centre is (2, 3)

<u>Method:</u> Using the formula above the equation of the circle is:

$(x - 2)^2 + (y - 3)^2 = 5^2$

We can re-write this as $x^2 - 4x + 4 + y^2 - 6y + 9 = 25$. Simplifying this we get:

$x^2 - 4x + y^2 - 4x - 6y + 13 = 25 \implies x^2 - 4x + y^2 - 4x - 6y - 12 = 0$

Example 2:

Find the centre of a circle and its radius if its equation is $(x - 3)^2 + (y + 4)^2 = 36$. Clearly the centre is 3, -4 and the radius is $\sqrt{36} = 6$ **Method**: Using the formula for the equation of a circle $(x - p)^2 + (y - q)^2 = r^2$ (where p, q are the centre co-ordinates and r is the radius)

Example 3:

Find the centre of a circle and its radius if the equation is $x^2 - 2x + y^2 - 4y - 11 = 0$

Method: Although this looks a bit tricky we need to make this equation similar to: $(x - p)^2 + (y - q)^2 = r^2$ first.

If we expand this we get $x^2 - 2px + p^2 + y^2 - 2yq + q^2 = r^2$

Equating this equation with the one given we get $-2p = -2$ which means p = 1

Likewise for the y term we get $-2q = -4$ which means q = 2

So the equation is $(x - 1)^2 + (y - 2)^2 = r^2$

Expanding the brackets we get $x^2 - 2x + 1 + y^2 - 4y + 4 - r^2 = 0$

$\implies$ $x^2 - 2x + y^2 - 4y + 5 - 16 = 0$ (we put -16 to adjust to make the equation the same as the original given. $(x-1)^2 + (y-2)^2 = 16$ Hence r = 4 and the centre of the circle is 1, 2

Practice Questions on equation of a circle

Find the equation of the circles (1 – 4) below:

(1) Centre (0, 2) radius 3

(2) Centre (1, -5) radius $\sqrt{2}$

(3) Centre (8, 15) radius 15

(4) Find the centre of a circle and its radius if its equation is $(x - 5)^2 + (y + 3)^2 = 49$

(5) Find the centre of a circle and its radius if the equation is $x^2 - 2x + y^2 - 4y - 44 = 0$

(6) PQ is a diameter of a circle. P is (-3, 6) and Q is (5, 12). Find the equation of the circle.

(7) The equation of a circle is given by $(x - 3)^2 + (y - 4)^2 = 49$

(a) Find the centre of the circle as well as the length of its diameter

(b) Sketch the circle on a graph

Answers to equation of circle questions

(1) $(x-0)^2 + (y-2)^2 = 9$ which simplifies to: $(x)^2+(y-2)^2 = 9$

We can re-write this as $x^2+(y^2 - 2y + 4) = 7 \Rightarrow$ $x^2+y^2 + 2y + 4 = 9 \Rightarrow x^2+y^2 + y - 5 = 0$

(2) $(x-1)^2 + (y+5)^2 = 2 \Rightarrow x^2+y^2 -2x + 10y +64 = 0$

(3) $(x-8)^2 + (y-15)^2 = 225 \Rightarrow x^2+y^2 -16x - 30y +64 = 0$

(4) Clearly the centre is 5, -3 and the radius is $\sqrt{49} = 7$
Method used: Using the formula for the equation of a circle $(x-p)^2 + (y-q)^2 = r^2$ (where p, q are the centre co-ordinates and r is the radius)

(5) The equation is $x^2 -2x + y^2 - 4y - 44 = 0$
Method:
Step 1: Equation of a circle can be represented as $(x-p)^2 + (y-q)^2 = r^2$
Step 2: If we expand this we get $x^2 - 2px + p^2 + y^2 - 2yq + q^2 = r^2$
Step 3: Equating this equation with the one given we get $-2p = -2$ which means $p = 1$. Likewise for the y term we get $-2q = -4$ which means $q = 2$. So the equation of the circle is $(x-1)^2 + (y-2)^2 = 49$

(6) Equation of the circle is $(x-1)^2 + (y-9)^2 = 25$

Method: First find the centre of the circle from the points of the diameter given. Clearly this is the mid-point of $PQ = \frac{-3+5}{2}, \frac{6+12}{2} = (1, 9)$. Now we need to find the length of the radius using Pythagoras's theorem. The length is $\sqrt{(1-5)^2 + (6+12)^2} = \sqrt{25} = 5$. So the equation of the circle is: $(x-1)^2 + (y-9)^2 = 25$

(7) Answer: Centre is (3, 4) and diameter 14 units

(7) **Method:**
The equation of a circle is given by $(x-a)^2 + (y-b)^2 = r^2$
Where the centre is (a, b) and radius is r units. This means in the equation given the centre is (3, 4) and radius 7 or diameter 14 units.

(b) Answer is shown by the sketch below

Method: We know the centre is at (3, 4) and the radius is 7 units hence the

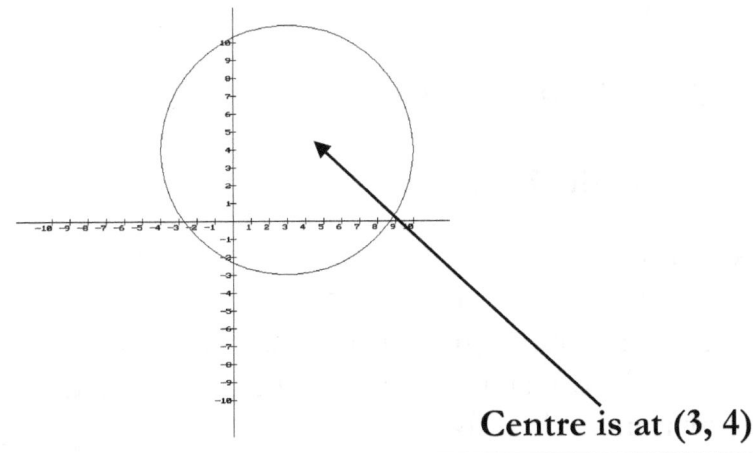

Centre is at (3, 4)

Algebra Section 9

Remainder and Factor Theorem

Remainder Theorem

If you have f(x) and you divide it by x – a the remainder will be f(a)

Example 1: Divide $f(x) = 4x^2 - 4x - 1$ by $g(x) = x - 1$

```
              4x, remainder - 1
        ┌─────────────────────
x - 1   │  4x² – 4x – 1
        │  4x² – 4x
        │  ─────────
        │   0   + 0 – 1
```

Check: Subsitute x= 1 in f(x) we get 4×1×1 - 4×1 – 1 = -1

Example 2: Divide $f(x) = 2x^3 - 3x^2 - 4x - 1$ by $x - 3$

Method: To find the remainder just work out f(3)

f(3) = 2×3×3×3 - 3×3×3 - 4×3 – 1 = 54 -27 -12 -1 = 14

Hence the remainder is 14.

Factor Theorem

If $f(a) = 0$, then this implies there is no remainder and x – a is a factor of the polynomial. Conversely, when x – a is a factor of a given polynomial then $f(a) = 0$

Example 1: Consider $x^2 - x - 2$ and investigate if x – 2 is a factor.

f(2) = $2^2 - 2 - 2 = 4 - 2 - 2 = 0 \implies$ x – 2 is a factor.

Example 2: Consider the cubic polynomial $4x^3 - 4x^2 - x + 1$

Show that 2x – 1, is a factor.

Method: If 2x – 1 is a factor of $4x^3 - 4x^2 - x + 1$, then $f(\frac{1}{2})$ should equal 0.

Let's test this by substituting x = $\frac{1}{2}$ in the cubic equation above.

We get $4(\frac{1}{2})^3 - 4(\frac{1}{2})^2 - \frac{1}{2} + 1 = \frac{4}{8} - \frac{4}{4} - \frac{1}{2} + 1 = \frac{1}{2} - 1 - \frac{1}{2} + 1$
= 0. Hence 2x -1 is a factor of $4x^3 - 4x^2 - x + 1$

The factor theorem can be very useful in finding one of the 'roots' of the equation. That is one possible solution to the equation. We can then try and find other factors by dividing the original cubic equation by 2x – 1 and factorise the expression we are left with. Although usually you don't have to divide as you can spot the other factors quickly.

See examples below in solving cubic equations.

Solving cubic equations

Example 1:

Solve the cubic equation $3x^3 + 4x^2 - 3x + 2 = 0$, given that one of its factors is x + 2

Step 1: Divide $3x^3 + 4x^2 - 3x + 2$ by x + 2

(We can divide this in the normal long division way as shown below)

$$
\begin{array}{r}
3x^2 - 2x + 1 \\
x + 2 \overline{\smash{)}3x^3 + 4x^2 - 3x + 2} \\
\underline{3x^3 + 6x^2 } \\
-2x^2 - 3x + 2 \\
\underline{-2x^2 - 4x } \\
x + 2 \\
\underline{x + 2 }
\end{array}
$$

So the cubic equation can be written as $(x + 2)(3x^2 - 2x + 1)$

In other words if one of the factors is x + 2, the other factor is $3x^2 - 2x + 1$

Example 2:

Solve the cubic equation $2x^3 + 3x^2 - 3x - 2 = 0$

This time we are not given a factor. In this case it is best to look at the constant term 2 and try x = 1, -1, 2 or -2 as possible solutions. (In other words factors of 2) Let's try x = -1 as a possible solution. If it is a solution then f(-1) should = 0.

Substituting f(-1) we get $2 \times 1^3 + 3 \times 1^2 - 3 \times 1 - 2 = 2 + 3 - 3 - 2 = 0$.

Hence we can re-write $2x^3 + 3x^2 - 3x - 2 = 0$ ****

As $(x - 1)(ax^2 + bx + c) = 0$

Clearly a = 2, since the only way to get $2x^3$ is to multiply x from the first bracket by $2x^2$ in the second bracket. Also we can figure out that c = 2. (Since -1 from the first bracket multiplied by c in the second bracket = -2 which means c = 2.

So far we have $(x - 1)(2x^2 + bx + 2) = 0$. Now multiply the 'x' terms out and equate with the x term in the expression ****

That is: 2x – bx = **-3x**. This means 5x = bx, hence b =5.

So the factors of $2x^3 + 3x^2 - 3x - 2$ are $(x - 1)(2x^2 + 5x + 2)$

We can now factor the quadratic expression $2x^2 + 5x + 2$ in the usual way into two brackets. That is $2x^2 + 5x + 2 = (2x +1)((x +2)$.

So finally the cubic equation $2x^3 + 3x^2 - 3x - 2 = 0$ can be written as: (x - 1)(2x +1)(x +2) =0   x = 1 or x =-$\frac{1}{2}$ or x = -2 (Note: We could have also factorised by dividing $2x^3 + 3x^2 - 3x - 2$ by x - 1 using long division shown earlier)

You can also use the graphical method for solving cubic equations as shown below:

Let $y = 2x^3 + 3x^2 - 3x - 2$, that is $f(x) = 2x^3 + 3x^2 - 3x - 2$. By choosing suitable values of x we can find corresponding values of y and plot the graph. The graph is shown below:

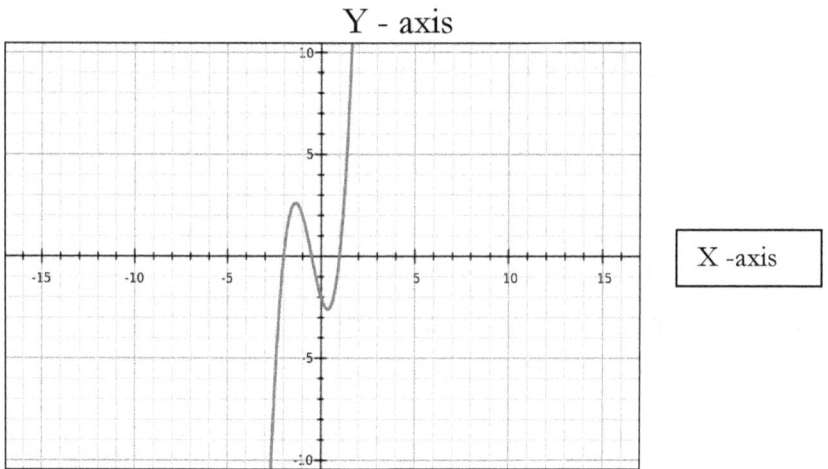

You can see from the graph that the roots are where the curve cuts the x –axis namely at x = –2, –0.5 and 1. Hence the roots of the equation $2x^3 + 3x^2 - 3x - 2$ are when x = –2, or x = –0.5 or x = 1

Practice Questions Algebra Section 9

(1) Solve the quadratic equation $6x^2 + x - 1 = 0$

(2) Find the intersection points where the line $y = x + 2$ meets the curve $y = x^2 - x + 1$

(3)(a) Using the remainder theorem show that when $f(x) = 2x^3 - x^2 - 18x + 12$ is divided by $x - 3$ the remainder is 3

(b) Prove that the remainder is 3, by dividing $f(x)$ by $x - 3$ using long division

(c) If $g(x) = (x - 1)(2x^3 - x^2 - 18x + 12)$ work out $g(-2)$

(d) Evaluate $fg(1)$

(4) A straight line given by the equation $y = 2x + 2$ intersects a curve $y = 2x^2 + 3x - 1$ at two points P and Q.

(a) Find the co-ordinates of these two points.

(b) Work out whether the curve $y = 2x^2 + 3x - 1$ has a maximum or minimum turning point.

(5) A parabola is given by the equation $y = x^2 + 2x - 8$. Write this equation in the form $y = (x + a)^2 + b$ and find the values of 'a' and 'b'.

Answers to Algebra Section 9

(1) Answer: $x = \frac{1}{3}$ or $x = \frac{-1}{2}$

Method: Factorise $6x^2 + x - 1 = 0$, to get $(3x - 1)(2x + 1) = 0$

$\Rightarrow$ $3x = 1$ or $2x = -1$ $\Rightarrow$ $x = \frac{1}{3}$ or $x = \frac{-1}{2}$

(2) Answer: $x = 1 + \sqrt{2}$ and $x = 1 - \sqrt{2}$ and $y = 3 + \sqrt{2}$ and $3 - \sqrt{2}$

Method: Solve the simultaneous equations (linear and quadratic)

$\Rightarrow$ $x + 2 = x^2 - x + 1$ $\Rightarrow$ $x^2 - 2x - 1 = 0$

$\Rightarrow$ Using the quadratic formula: $x = \frac{-b \pm \sqrt{b^2 - 4ac}}{2a}$

$\Rightarrow$ $x = \frac{2 \pm \sqrt{2 \times 2 - 4 \times 1 \times (-1)}}{2 \times 1} = \frac{2 \pm \sqrt{4+4}}{2} = \frac{2 \pm \sqrt{8}}{2} = \frac{2 \pm 2\sqrt{2}}{2}$

$= 1 \pm \sqrt{2}$, then substitute for x to find corresponding values of y

(3) (a) Answer: 3

Method: Substitute $x = 3$ in $f(x)$ $\Rightarrow$ $f(x) = 2 \times 27 - 9 - 54 + 12 = 3$

(b) Answer: See long division below

Method: Do normal long division

$$\require{enclose}
\begin{array}{r}
2x^2 + 5x - 3 \text{ remainder } 3 \\
x - 3 \enclose{longdiv}{2x^3 - x^2 - 18x + 12} \\
\underline{2x^3 - 6x^2} \\
5x^2 - 18x \\
\underline{5x^2 - 15x} \\
-3x + 12 \\
\underline{-3x + 9} \\
3
\end{array}$$

(c) Answer = -84

Method: work out g(-2) by substituting x = -2 in equation g(x) ⟹ (-2 -1)(2×-8 - 4 + 36 + 12) = -3(-16 -4 +48) = -3(28). **Answer = -84**

(d) Answer fg(1)) = 12

Method: First work out g(1). This =0 ⟹ f(0) = 12 Hence fg(1) = 12

(4) (a) **Answer: (-1.5, -1) and (1, 4)**

Method: Solve the two equations above:

$\implies 2x^2 + 3x - 1 = 2x + 2 \implies 2x^2 + x - 3 = 0$

$\implies (2x + 3)(x - 1) = 0 \implies x = -1.5 \text{ or } 1$

Substituting for x in the equation y = 2x + 2, we can find the co-ordinates of the two intersecting points.

(b) **Answer: Minimum**
   Method: Sketch the graph roughly and you will see it is a 'U' shaped curve with a minimum

   (5) a = 1 and b = -9

Algebra Section 10

Linear equations

These are of the form $y = mx + c$

where **m** is the **gradient or slope** and c is the value of y when $x = 0$

Example 1: $y = 3x - 1$

The graph is shown below.

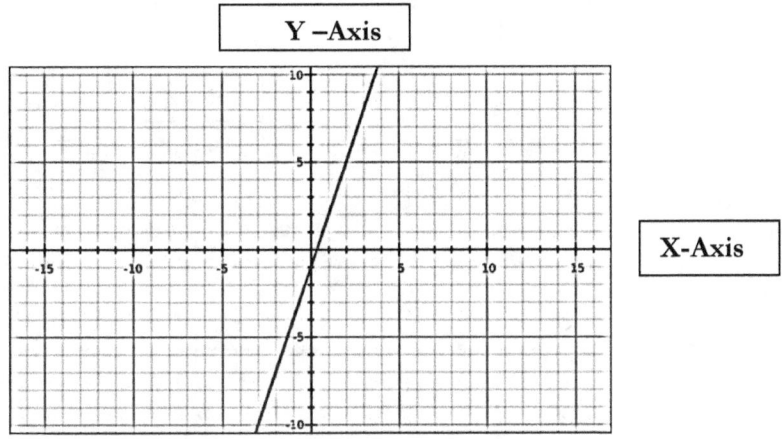

You can see that the equation $y = 3x - 1$ crosses the y–axis at $y = -1$ (this is the called the intercept)

In other words when $x = 0$, $y = 3 \times 0 - 1 = -1$

The '3' in the 3x bit refers to the gradient or the slope of the graph.

So in general a linear equation is of the form y = mx + c, where m is the gradient and c is the value of y when x = 0

Example 1

Plot the equation y = 2x - 3 for values of x = -2 to +2 by completing the table below first. The plotted graph is shown below the completed table.

x	-2	-1	0	1	2
2x – 3	-7	-5	-3	-1	1
y	-7	-5	-3	-1	1

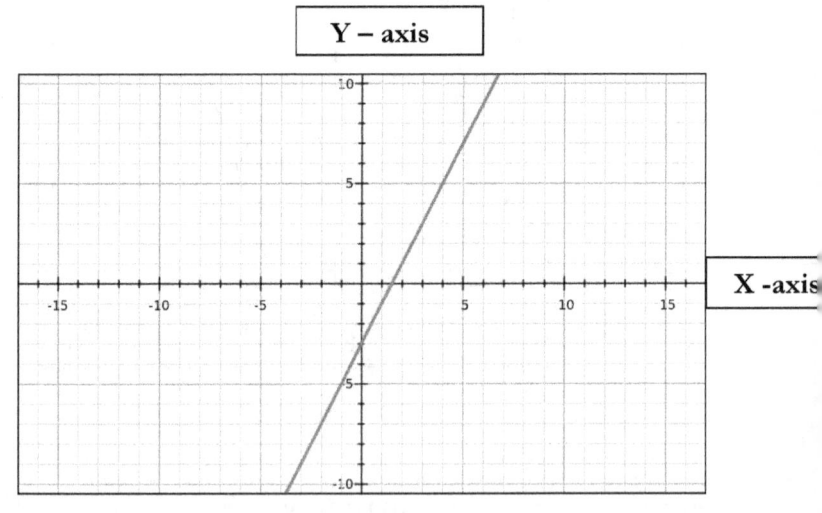

Example 2

The co-ordinates P (2, 3) and Q (4, 6) lie on a straight line.

(i) Find the mid-point R of the co-ordinates P & Q

(ii) Find the gradient of the straight line

(iii) Find the equation of the straight line

(iv) Find the distance between the two points P & Q

(i) The mid-points are simply the average of the x co-ordinates and the y co-ordinates.

Mid –point for the x co-ordinate $= \frac{x_1+x_2}{2} = \frac{2+4}{2} = 3$

Similarly, the mid-point of the y co-ordinate $= \frac{y_1+y_2}{2} = \frac{3+6}{2} = 4.5$

Hence R, the mid-point of PQ is (3, 4.5)

(ii) The gradient of two points that lie on a straight line $= \frac{Difference\ in\ y\ co-ordinates}{Difference\ in\ x\ co-ordinates} = \frac{6-3}{4-2} = \frac{3}{2} = 1.5$

(iii) (a) The equation of a straight line is given by y =mx + c

Since m = 1.5, then the equation is y = 1.5x + c
We also know that it goes through P, Q and R.
So we can choose any of these to find the value

of C. Let us choose P(2, 3) so the equation is now:

3 = 1.5×2 + C, Hence 3 = 3 + C so in this case C =0

So the equation is y = 1.5x

(b) You can also find the equation of a straight line if you know the co-ordinates of a point it goes through and its gradient by using the formula: y − y1 = m(x − x1). (where x1, y1 are the co-ordinates of the point the line goes through and m is the gradient).

(iv) To find the distance between P & Q we simply use Pythagoras's theorem and find that:

$$PQ = \sqrt{(4-2)^2 + (6-3)^2} = \sqrt{4+9} = \sqrt{13}$$ units

Using Ratios to find Co-ordinates

Example: You are given that M(-2, 2) and N(3, 12) are end points of a line. Also that P lies on this line such that MP: PN = 3:2. Find the co-ordinates of P.

Method:

Step 1: Find the difference in x and y co-ordinates

Difference in x = 3 – (-2) = 5

Difference in y = 12 – 2 = 10

Step 2: We know that P lies on the line MN in the ratio 3:2

This means P is $\frac{3}{5}$ of the way up from M

Step 3: To find P we use this ratio so that x = $\frac{3}{5} \times 5 = 3$ and y = is $\frac{3}{5} \times 10 = 6$

Step 4: Finally, to find P we simply add the co-ordinates of M to the co-ordinates above. So P = (-2 + 3, 2 + 6) = (1, 8)

Working out equations of 'Normals' and 'Parallel' lines

If two lines are perpendicular to each other, then their gradients, m1 & m2 when multiplied together = −1. That is $m_1 m_2 = -1$ (The line that is perpendicular to given line is called a 'normal')

Example 1

Given the equation $y = 3x - 1$. Find the equation of a line that is perpendicular to it and goes through the co-ordinates P(1, -2)

Method: Gradient of the line $y = 3x - 1$ is 3 (since if $y = mx + c$, then m is the gradient). Clearly if $m1 \times m2 = -1$ then $3 \times m2 = -1$. This means $m2 = -\frac{1}{3}$. Finally if this line goes through P (1, -2) then its equation can be found by using the fact that $y - y1 = m(x - x1)$ where x1, y1 are the co-ordinates it goes through. This means the equation of the **normal** or the **line that is perpendicular** to $y = 3x - 1$ is given by $y - (-2) = -\frac{1}{3}(x - 1)$. (We simply substituted the co-ordinates of P(1,-2) in the equation $y = 3x - 1$)

$\implies y + 2 = -\frac{1}{3}(x - 1) \implies 3y + 6 = -x + 1 \implies 3y + x = -5$

$\Rightarrow$ $y = -\frac{x}{3} - \frac{5}{3}$ or $y = -\frac{x}{3} - 1\frac{2}{3}$

Alternative method for finding the perpendicular line:

Since the equation of a straight line is given by $y = mx + c$, then the equation of the perpendicular line is $y = -\frac{1}{3}x + c$

We can find c by substituting P(1, -2) in this equation. Substituting for x and y we get: $-2 = -\frac{1}{3} \times 1 + c$. This means $c = -2 + \frac{1}{3} = -1\frac{2}{3}$; hence $y = -\frac{1}{3}x - 1\frac{2}{3}$

Finding parallel lines

This is more straightforward **since the gradient of parallel lines are the same.**

Example: Find the line that is parallel to $y = \frac{1}{2}x + 3$ which passes through the point (2, -1)

Method: We know that the equation of a line is given by $y = mx + c$

Since the gradient of this parallel line is the same, the equation of this parallel line is: $2 = \frac{1}{2} \times (-1) + c \Rightarrow c = 2 + \frac{1}{2} = 2\frac{1}{2}$.

So the equation of the **parallel line** which goes through (2, -1) is $y = \frac{1}{2}x + 2\frac{1}{2}$

Algebra section 11

Logarithms

Some useful definitions when using logarithms

A logarithm (log) is the inverse of exponentiation

$b^y = x$

This means $\log_b x = y$

$\log_b x$ is called the logarithm of y to base b

For example if $2^3 = 8$, using the definition of logarithm we can write

$\log_2 (8) = 3$

Notice that logarithm is an inverse function of an exponential function

For example if: $y = \log_a(x)$

Then this is the inverse function of the exponential function,

$x = a^y$

Inverse logarithm calculation

The inverse logarithm (or anti logarithm) is calculated by raising the base b to the logarithm y:

$x = \log^{-1}(y) = b^y$

Logarithmic function

The logarithmic function has the basic form of:

$f(x) = \log_b(x)$

<u>Important logarithm rules to remember:</u>

Logarithm Rules

Product Rule: $\log_b(x \cdot y) = \log_b(x) + \log_b(y)$
The logarithm of the product of x and y is the sum of the logarithms of *x* and *y*. **Use this rule to compute the logarithm of a x.y, when you know the logarithm of x and the logarithm of y.**

Quotient Rule: $\log_b(x/y) = \log_b(x) - \log_b(y)$
The logarithm of the quotient of x and y is the difference of the logarithms of x and y. **Use this rule to compute the logarithm of x/y, when you know the logarithm of x and the logarithm of y.**

Power Rule: $\log_b(x^y) = y \cdot \log_b(x)$
The logarithm of (x raised to the power of y) is the product of y and the logarithm of x. **Use this rule to compute the logarithm of x^y, when you know the logarithm of x.**

Power of Base Rule: $\log_b(b^y) = y$
This is just a statement of the fact that logarithm is the inverse of exponentiation. It can also be considered a special case of the Power Rule (it follows from the Power Rule combined with the Logarithm of the Base rule below).
Use this rule to compute the logarithm of a power of the base.

Base Change Rule: $\log_b(x) = \log_c(x)/\log_c(b)$
The logarithm of x to base b is the quotient of the logarithm of x to base c and the logarithm of b to base c.
Use this rule to compute the logarithm of x to base b, when you know the logarithm of x to a different base (c).

Logarithm of the Base: $\log_b(b) = 1$
The logarithm of b to base b is always 1.

Logarithm of 1: $\log_b(1) = 0$
The logarithm of 1 to any base is 0.

Examples:

(1) $\log_{10}(64) = \log_{10}(4 \times 16)$ (because $64 = 4 \times 16$)
 $= \log_{10}(4) + \log_{10}(16)$ **(Product Rule)**

(2) $\log_{10}(0.5) = \log_{10}(1/2) = \log_{10}(1) - \log_{10}(2)$ **(Quotient Rule)** but $\log 10(1) = 0$ **(Logarithm of 1 to any base is 0)**
So, $\log_{10}(0.5) = - \log_{10}(2)$

(3) $\log_{10}(64) = \log_{10}(2^6)$ (because $2^6=64$) $= 6.\log_{10}(2)$ **(Power Rule)**

(4) $\log_{10}(64) = \log_2(64) / \log_2(10)$ **(Base Change Rule)**

$= 6/\log_2(10)$ (because $2^6=64$)

Some more examples

Logarithm product rule

As we saw earlier in the product rule the logarithm of the multiplication of x and y is the sum of logarithm of x and logarithm of y.

$log_a(x \cdot y) = log_a(x) + log_a(y)$

For example:

$log_{10}(4 \cdot 8) = log_{10}(4) + log_{10}(8)$

Logarithm quotient rule

Similarly, the logarithm of the division of x and y is the difference of logarithm of x and logarithm of y.

$log_a(x / y) = log_a(x) - log_a(y)$

So for example:

$log_{10}(3 / 7) = log_{10}(3) - log_{10}(7)$

Logarithm power rule

The logarithm of x raised to the power of y is y times the logarithm of x.

$log_a(x^y) = y \cdot log_a(x)$

For example:

$log_{10}(2^6) = 6 log_{10}(2)$

Logarithm of infinity

The limit of the base b logarithm of x, when x approaches infinity, is equal to infinity:

$\lim \log_b(x) = \infty$, when $x \rightarrow \infty$

Example 1:

Find x in the log equation below:
$\log_2(x-1) + \log_2(x+1) = 3$

Solution:
$\log_2((x - 1).(x+1)) = \log_2(x-1) + \log_2(x + 1)$ **(Product Rule)**

$=> \log_2((x - 1).(x+1)) = 3$

$=> (x -1).(x + 1) = 2^3$ (from the definition of logarithm)

$=> x^2 - 1 = 8$

$=> x^2 = 9$

$=> x = 3$

($x = -3$ would also satisfy $x^2 = 9$ but plugging -3 back into the original problem statement would lead to the logarithm of a negative number, which is not allowed).

Example 2:

Find x for

$\log_3(x+2) - \log_3(x) = 2$

Solution:

Using the quotient rule:

$\log_2((x+2)/x) = 2$

Changing the logarithm form to the exponential form:

$(x+2)/x = 2^2$

Which simplifies to:

$x+2 = 4x$

Or

$3x = 2$

Or

$x = \dfrac{2}{3}$

Practice Questions on Logarithms

(1) Re-write $3^4 = 81$, in logarithmic form:

(2) Compute $\log_{10}(20) + \log_{10}(5)$

(3) Compute $\log_{10}(0.1)$

(4) Simplify $\log_{10}(7/8)$

(5) Find x given $\log_3(x+4) - \log_3(2x) = 2$

(6) Find x in the expression $\log_2 x + \log_2 10 = 6$

Answers to logarithm questions

(1) Answer: $\log_3(81) = 4$

(2) Answer: $\log_{10}(100) = 2$ (Using the product rule and the power of base rule)

(3) Answer: -1 (Using the quotient and the logarithm of 1 rule)

(4) Answer: $\log_{10}(7) - \log_{10}(8)$

(5) Answer: $x = \frac{4}{15}$, since $(x+4)/2x = 2^3$ so $x + 4 = 2x \times 8$, $x + 4 = 16x$, hence $15x = 4$ or $x = \frac{4}{15}$

(6) Answer: $10x = 2^6$ which means $10x = 64$ so $x = 6.4$

Algebra Section 12

Sequences

Arithmetic sequences:

Working out a general formula for an arithmetic sequence:

You can either use the difference method or the general formula to find the nth term of an arithmetical sequence.

Example 1: (a) Find the nth term of the sequence 5, 8, 11, 14, ----- **(b)** Hence find the 15th term.

Method: The common difference in this case is 3. So we multiply the common difference by n to get 3n. However each term is 2 more than 3n. (a) Hence the nth term is 3n + 2

(b) To find the 15th term simply substitute n = 15 in 3n +2. Hence the 15th term is $3 \times 15 + 2 = 45 + 2 = 47$

Example 2: Find the nth term of the sequence 4, 1, -2, -5

Method: This time the common difference is -3. So the nth term is $7 - 3n$

(Since each term is 7 more than -3n so the nth term is $7 - 3n$

This time consider a general arithmetical sequence as shown:

a, a +d, a+2d, a +3d, a+4d, a+5d, a+6d, ... (d is the common difference). We can see that the second term is a+d, the third term is a+2d, the fourth term is a+3d, the fifth term is a+4d or a + (5-1)d, the sixth term is a+5d or a +(6-1)d

The seventh term is a+6d or a + (7- 1)d, so the nth term is a+(n-1)d

You can check to see if this is right by substituting n=1, 2, 3, 4, 5 and so on to the appropriate numbers in the sequence. See example below:

Example 2: Find the nth term of the arithmetical sequence below:

5, 9, 13, 17 ... this is an arithmetical or linear sequence since the numbers go up by the same constant number. We know the nth term is a + (n - 1)d

In this case a=5 (This is the first term) d = 4 (this is the common difference between each successive number). So, the nth term is 5 +(n – 1)× 4 = 5 +4n –4 = 4n +1,

Quadratic Sequence

A quadratic sequence is a sequence of numbers in which the second difference between any two consecutive terms is constant.

Consider the following example: 6, 11, 18, 27, 38…

The first difference is calculated by finding the difference between consecutive terms:

So we get 5, 7, 9, 11…..

The second difference is obtained by taking the difference between consecutive first differences:

This time we get 2, 2, 2, ……

We notice that the second differences are all equal to 2. **Any sequence that has the same common second difference is a *quadratic sequence.***

So we now know the first term is n^2 we now need to find the remaining linear sequence to give us 6, 11, 18, 27, 38, …….

We subtract n^2 (n =1, 2, 3, 4, ….) from this sequence: 6 – 1, 11 – 4, 18 – 9, 27 -16, 38 – 25 to give us 5, 7, 9, 11, 13, …… We can see this is a linear sequence 2n + 3.

Hence the quadratic sequence required is n^2 + 2n + 3

Some other sequences you should be familiar with

Cube numbers

Similarly 1, 8, 27, 64, ...is the cube of natural numbers. The nth term is n^3

Fibonacci Series

1, 1, 2, 3, 5, 8, 13, 21 ......   (To get the next number, add the previous two) e.g. 1 +1 = 2, 1+2 =3, 2+3 = 5, and so on.

Multiply or divide each number by the same number

(1) 2, 6, 18, 54 ….. each number is 3× the previous number
(2) 12, 6, 3, 1.5 …each number the previous number divided by 2

Recurrence Relations

A recurrence relation gives you a connection between two consecutive terms and can help you to work out a term in a sequence from the previous term.

Consider first the notation that can be used such as a_n, a_{n+1}, a_{n-1}

a_1 means the first term

a_2 means the second term

a_n means the nth term

Example 1:

If we say $a_{n+1} = a_n + 5$, this means the next term in the sequence is 5 more than the previous term. So if $a_n = 19$ then $a_{n+1} = 19 + 5 = 24$

Example 2:

Consider the sequence 1, 5, 9, 13. ......

This could be written as a recurrence relation as $a_{n+1} = a_n + 4$, also $a_1 = 1$

If $a_1 = 1$ (the first term) then clearly, $a_2 = 1 + 4 = 5$ and so on.

Example 3: Find the recurrence relation of the sequence 4, 11, 18, 25, 32, ...

We can see that $a_{n+1} = a_n + 7$, and $a_1 = 4$

Example 4: A sequence is given by the recurrence relation $a_{n+1} = a_n - 3$ and $a_1 = 4$. Find the first five terms.

The second term $a_2 = 4 - 3 = 1$

The third term $a_3 = 1 - 3 = -2$

$a_4 = -2 - 3 = -5$ and $a_5 = -5 - 3 = -8$

Hence the first five terms are 4, 1, -2, -5 and -8

Example 5: A recurrence relation generates a sequence given by $a_{n+1} = 2a_n + R$

You are also given that $a_1 = 3$

(a) Find a_2 in terms of R
(b) Given that $a_3 = 18$, find the value of R
(c) Hence find the general recurrence relation

Method: Using the initial relation we get, $a_2 = 2a_1 + R$

Substituting the values we get $a_2 = 2 \times 3_1 + R$ (Since we are given that $a_1 = 3$)

Hence $a_2 = 6 + R$. Similarly, $a_3 = 2a_2 + R$

$\implies a_3 = 2(6 + R) + R$

$\implies a_3 = 12 + 2R + R \implies a_3 = 12 + 3R$

$\implies 18 = 12 + 3R$ (Since we are given that $a_3 = 18$)

$\Rightarrow 3R = 6 \Rightarrow R = 2$

Hence $a_2 = 2a_1 + 3$ or more generally $a_{n+1} = 2a_n + 3$

Fractional sequences

Find the 3rd term of the sequence $\dfrac{n}{3n-1}$

We simply substitute n = 3 in the expression above to give us $\dfrac{3}{3\times 3 - 1} = \dfrac{3}{8}$

Limiting value of a sequence

Find the limiting value of $\dfrac{3n+4}{6n-3}$ as n $\longrightarrow \infty$

Initially we simply divide each term in the numerator and denominator by n to get: $\dfrac{3+\frac{4}{n}}{6-\frac{3}{n}}$. We now can see that as n tends to ∞, the expression $\dfrac{3+\frac{4}{n}}{6-\frac{3}{n}} = \dfrac{3}{6} = \dfrac{1}{2}$

Finding the sum of natural numbers:

To find the sum of natural numbers: 1, 2, 3, 4, 5, ------------ n

To do this use the formula we just found where a = 1 and d = 1 hence

$$S_n = \frac{n}{2} \times (2 + (n-1) \times 1), \quad \text{Hence } S_n = \frac{n}{2}(2 + (n-1))$$

$$= \frac{n}{2}(2 + n-1) = \frac{n}{2}(1 + n)$$

So we get: $S_n = \frac{n(n+1)}{2}$

Note: We normally use the symbol Σ (Sigma) to mean the sum of a series

(Σ is a Greek letter sigma which stands for S)

Example 1: Find the sum of the first 50 natural numbers.

Method: Substitute n = 50 in the formula $S_n = \frac{n(n+1)}{2}$,

Hence, $S_n = \frac{50(50+1)}{2} = 25 \times 51 = 1275$

So the sum of the first 50 natural numbers is **1275**

Example 2: An arithmetic series has the following sequence: 3, 6, 9, 12, 15,……. Find the sum of the first 20 terms.

Method:

First find the 20th term using the formula a + (n – 1)d ⇒
20th term = 3 + (20 –1)×3 = 60

Now, using the formula $S = \frac{n}{2} \times (a + l)$ substitute a = 3 (the first term) and l =60 (**the last term**)

⇒ $S = \frac{n}{2} \times (a + l) = \frac{20}{2} \times (3 + 60) = 10 \times 63 = 630$
⇒ the sum of the first 20 terms for the given arithmetic sequence is **630.**

Geometric Sequences and Sums:

A Geometric Sequence is found by multiplying the previous term by a constant.

Example 1:

    3, 9, 27, 81, 243, …….

In this case each term (except the first term) is found by **multiplying** the previous term by 3.

In General we can write a Geometric Sequence like this:

$\{a, ar, ar^2, ar^3, ... \}$

Where **a** is the first term, and **r** is the factor between the terms (called the **"common ratio"**)

Finding the **nth** term:

We can also calculate the nth **term** using the Rule:

$$x_n = ar^{(n-1)}$$

(Note that we use "n - 1" because ar^0 is for the 1st term)

Example: Find the 9th term of the geometric sequence:

5, 10, 20, 40, 80, 160, ...

This sequence has a factor of 2 between each number.

The values of **a** and **r** are: **a = 5** (the first term) and **r = 2** (the "common ratio")

So the nth term is given by: $x_n = 5 \times 2^{(n-1)}$

So, **nth** term is 9:

$$X_5 = 5 \times 2^{(9-1)} = 5 \times 2^8 = 5 \times 256 = 1280$$

A Geometric Sequence can also have **smaller or fractional** values:

Example: 2, 1, 0.5, 0.25, ...

This sequence has a factor of 0.5 between each number.

So to find the nth term we simply use: $x_n = 4 \times (0.5)^{n-1}$

Summing a Geometric Series

We can sum a Geometric Sequence, using the formula below:.

To sum:

$$a + ar + ar^2 + ... + ar^{(n-1)}$$

Each term is ar^k, where k starts at 0 and goes up to $n - 1$

Using the formula below we have:

$$\sum_{k=0}^{n-1} (ar^k) = a \left(\frac{1 - r^n}{1 - r} \right)$$

where a is the first term
r is the "**common ratio**" between terms
and **n** is the number of terms
"

You have already seen the Σ Notation simply means 'to sum up'

Example: Sum the first 6 terms of : 5, 10, 20, 40, 80, 160,

The values of **a, r** and **n** are:

- a = 5 (the first term)
- r = 2 (the "common ratio")
- n = 6 (we want to sum the first 6 terms)

So:

$$\sum_{k=0}^{n-1} (ar^k) = a \left(\frac{1-r^n}{1-r} \right)$$

Using the formula we get the answer to be 315.

As you can see although this is an easy example if you had to sum up to 25 numbers or more the formula would be very useful.

Practice Questions on Sequences

(1) Find the nth term of the linear sequence
-8, -3, 2, 7, 12, ...

(2) Find the 8th term of the sequence $\dfrac{5n}{2n-6}$

(3) Find the limiting value of $\dfrac{5n-1}{8n+3}$ as n ⟶ ∞

(4) Find the nth term of the quadratic sequence: 2, 3, 6, 11, 18, ...

(5) Find the nth term of the quadratic sequence:

5, 11, 19, 29, 41.......

(6) Find the limiting value of $\dfrac{4n+1}{7n+2}$ as n ⟶ ∞

(7) Find the 10th term of the sequence $\dfrac{7n}{22n+1}$

(8) Find the sum of the first 10 terms of the arithmetic sequence: 2, 5, 8, 11, 14......

Answers to Questions on Sequences

(1)   5n – 13

(2)   4

(3) $\dfrac{5}{8}$

(4) $n^2 - 2n + 3$

(5) $n^2 + 3n + 1$

(6) $\dfrac{4}{7}$

(7) $\dfrac{70}{221}$

(8)   155 - Remember first find the last term in this case the 10th term, then use the formula $S = \dfrac{n}{2} \times (a + l)$

Algebra Section 13

Sets and Venn Diagrams

> **Sets**
>
> A <u>set</u> is simply a collection of things.
>
> **Example 1**: Consider all the numbers from to 2 to 7 inclusive. These are 2, 3, 4, 5, 6 and 7. Assume we can represent these numbers by set A then we can write this as A : { 2, 3, 4, 5, 6, 7)
>
> **Example 2**: The pupils who got top grade in English in 2012 in a certain school were Jill, Fatima and John. Assume we represent this by set B.
>
> We can write this as B: {Jill, Fatima, John}
>
> **Example 3**: C represents the set of square numbers. We can write these square numbers as C: {1, 4, 9, 16, 25, 36, ...}
>
> **Each item, name, number, etc is referred to as a member of the set.**
>
> **You can also represent these same sets by circles in a Venn diagram.**
>
> **The rectangle outside the circle(s) is called the Universal set and contains all the members in the rectangle.**

Symbols and notation associated with Sets:

You need to be familiar with some basic symbols associated with sets. For example 'U' stands for Union, so AUB means it belongs either to set A or B when two sets are involved; Or in all three sets A or B or C when three sets are involved.

The symbol ∩ stands for Intersection, so that A∩B means it must be in both sets A and B (this is the overlapping part of set A and set B), in the case of two sets and in all three sets A and B and C when three sets are involved. This will be clearer in the examples shown a bit later which illustrate these points.

Also A ⊆ B means A is a subset of B. This means set A is part of set B. Also note that A' means the complement of A that is all the elements that do not belong to set A.

U refers to the universal set. This is all the elements in the rectangle.

∅ This symbol denotes the empty set - a set with no items or elements in it

Element of a set

ϵ means it is an **element** of a set and ϵ means it is **not an element** of a set

Also n(ϵ) means the number of elements in a given set

Basics of Venn Diagrams

A Venn diagram is a pictorial way to represent sets. As mentioned before, the rectangle outside the set is called the Universal set normally represented by U.

You already know that Set A is simply all the members of Set A and likewise Set B is all the members of Set B. Now let us consider some examples of Venn diagrams involving just two sets.

Example 1:

A∪B is simply all of A or all of B as represented by the shaded diagram below:

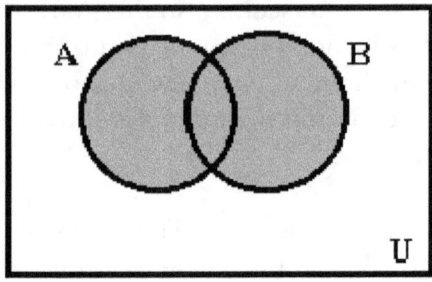

Example 2:

Similarly A ∩ *B* (A intersection B) is the dark shaded bit, that is where the two circles intersect.

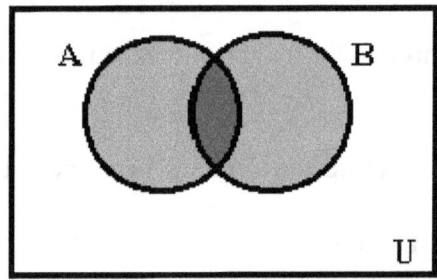

Example 3:

A' or the complement of A that is everything in the Universal set but not in A as shown by the shaded bit below:

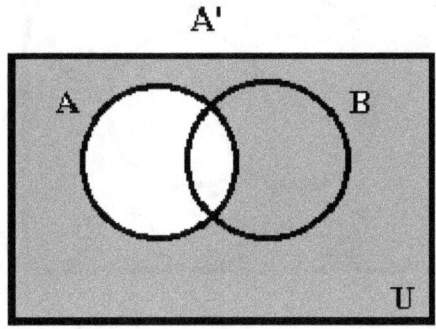

Example 4: Now consider (AUB)'

First shade what's in the brackets, **namely A U B (A union B)**

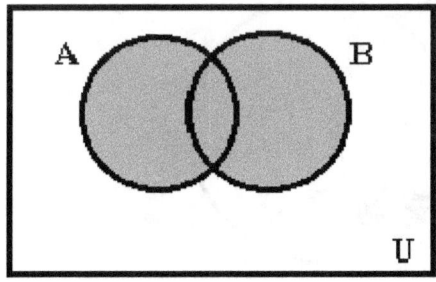

Now take the complement of the union of these sets. This is shown by the shaded bit in the diagram below:

(A∪B)'

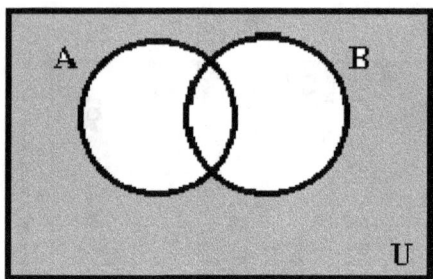

Venn Diagrams for 3 Sets

Unions, intersections and Complements, are handled in the same way as they are with 2 set Venn Diagrams. For example the intersection of A∩B∩C is shown by the shaded bit where the three circles overlap.

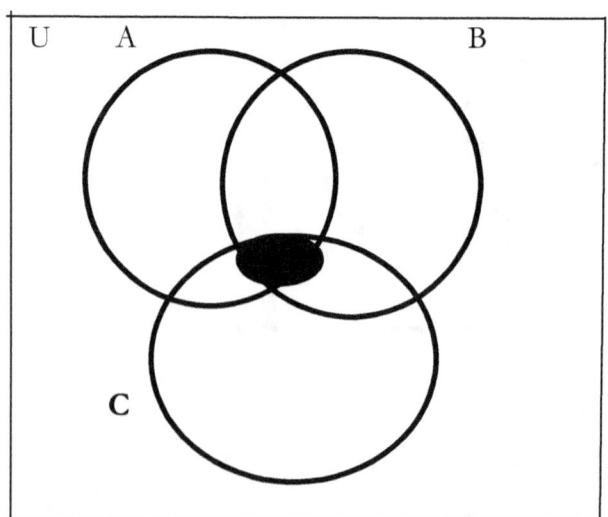

Examples: Let us start initially with simple questions involving just one set. Assume all consecutive numbers from 2 to 7 can be represented by the Universal set U and that Set A consists of odd numbers in this Universal set. Show this in a Venn diagram.

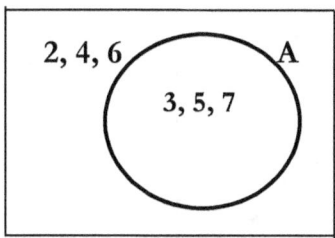

Explanation: The rectangle represents the Universal set. All the numbers in the circle set A are odd numbers, namely 3, 5 & 7. The remaining numbers 2, 4 & 6 are outside the circle.

Example 2: The pupils who got top grade in English in 2012 in a certain school were Jill, Fatima and John. Assume we represent this by set B. This is shown in the Venn diagram below. The rectangle represents the Universal set and B represents those students who got top grade in English.

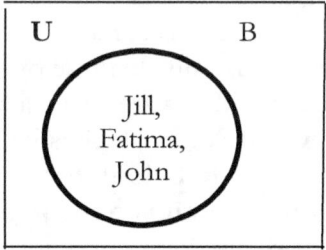

Now let us consider typical questions you are likely to get in the Exam.

Example 1: U is the universal set. A represents the set of odd numbers and B represents the set of square numbers, show how the number, 1, 4, 9, 12 and 16 can be illustrated by a Venn diagram.

The Venn diagram that shows the above information is shown below:

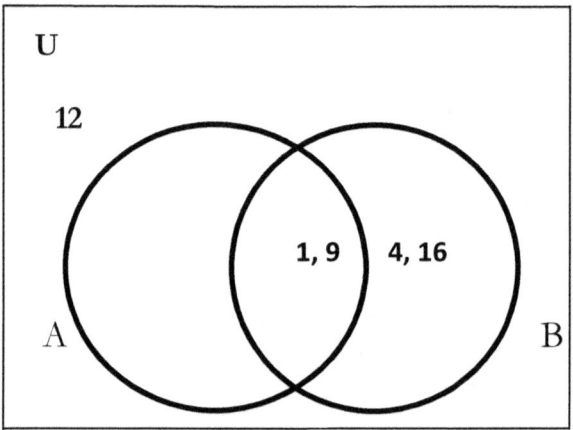

Explanation: The rectangle represents the Universal set. **Circle A** represent the **set of odd numbers** and **circle B represent the set of square numbers**. The numbers 1, 4, 9, 12 and 16 have to be fitted inside the Venn diagram such that the conditions stated are satisfied. Namely, the odd numbers are in set A and the square numbers are in set B and any other number is outside these two sets. You probably noticed that 1 & 9 are not only odd but are also square numbers. **Hence 1 & 9 belong to both Set A and Set B** which is inside the intersection of the two circles. **Clearly 4 & 16 belong to the part of B as shown. Finally, 12 is an even number which is outside both set A and B as shown.** As stated before 1 & 9 belong to the intersection of A and B or more formally in **A∩B** .

Example 2: The circles in the Venn diagram below represent passes with exams in Science, Maths and English. Janet (J) passes in all these three subjects. Alex (A) passes in English and Math but fails to get this in Science. Represent this information by a Venn diagram.

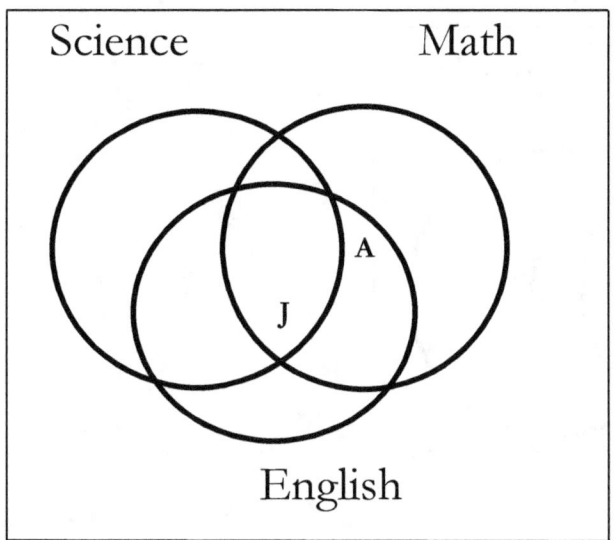

Explanation: The Venn diagram above represents Janet's (J) and Alex's (A) position as shown.

Example 3: A represents the set of even numbers and set B represents multiples of 5, show how the numbers 8 to 15 inclusive can be illustrated by a Venn diagram:

The answer is shown below:

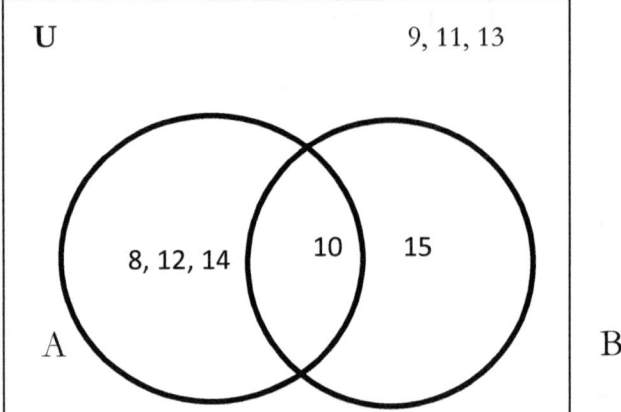

Practice Questions on Sets

(1) Consider the Venn diagram below:

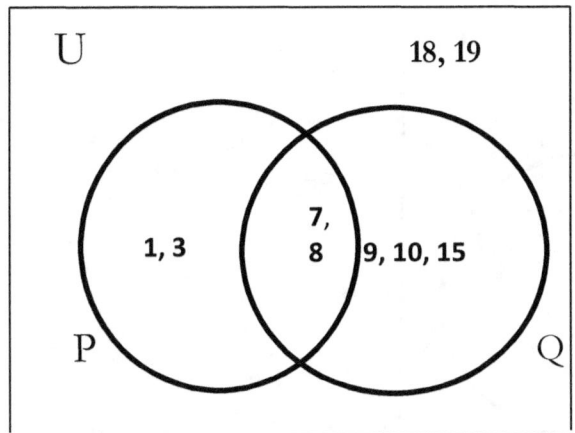

Complete the following by listing the numbers inside the brackets. The first one is done for you.

(i) P = {1, 3, 7, 8}

(ii) Q = {   }

(iii) P U Q = {   }

(iv) P ∩ Q = {   }

(v) U = {   } (**Remember U corresponds to the Universal set**)

(2) In a certain year group 50 pupils took an exam in English or Math a year earlier than normal. The data is shown in the Venn diagram below where 'x' represents the number of students who took both Math and English.

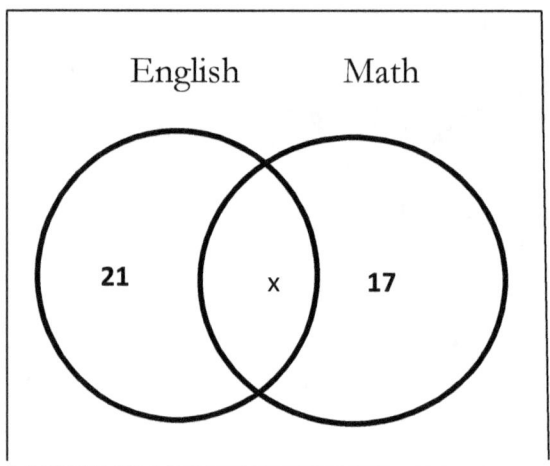

(i) How many students take both English and Math (represented by x)?

(ii) How many students do not take English?

(iii) How many students just take Math?

(3) Consider the Venn diagram below and answer the questions that follow:

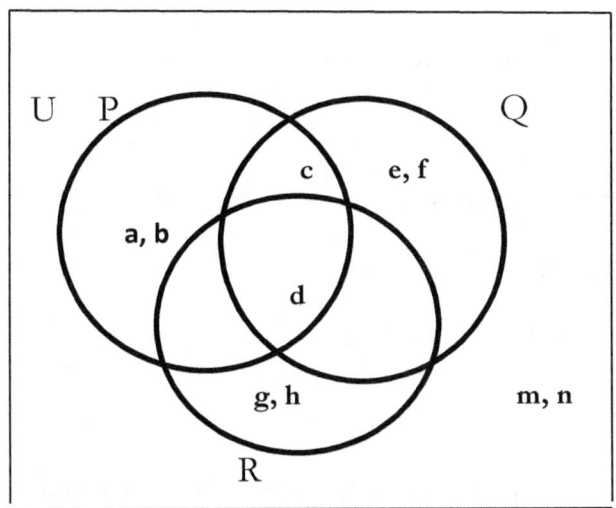

(i) n(E) =? (How many elements are there in the Universal set?)

(ii) P U Q = {   }

(iii) P ∩ Q ∩ R = {   }

(iv) P U Q U R' = {   }

(v) P' ∩ Q U R = {   }

(4) Assume A is the set of factors of 12, B is the set of prime numbers less than 11 and C is the set of even numbers less than 8. What numbers belong to (A∪B)∩C?

Click/circle the correct answer:

(a) {1,2,4}   (b) {1,2,6}   (c) {2,3,12}   (d) {2,4, 6}

Answers to Sets:

(1) (i) P = { 1, 3, 7, 8}

Explanation: This is simply all the elements (numbers) in circle P

(ii) Q = {7, 8, 9, 10, 15}

Explanation: Similarly, all the numbers in Q

(iii) P∪Q = {1, 3, 7, 8, 9, 10, 15}

Explanation: Numbers that belong to P OR Q

(iv) P∩Q = {7, 8}

Explanation: Numbers that belong to both P AND Q (that is the overlapping numbers)

(v) U = {1,3,7,8,9,10,15,18,19}

Explanation: All the numbers in the Universal set, that is in the rectangle

(2) (i) 12

Explanation: The number of students that take both English and Maths is found by working out x (the number in the overlapping or intersecting circles). We know the total number of students is 50. This means 21 + x + 17 =50 which simplifies to 38 + x =50. So x= 12. Hence, 12 students belong to the intersecting circles A and B

(ii) 17

Explanation: The number of students that do not take English is 17. This is found by excluding the total in the Maths circle.

(iii) 17

Explanation: This is really the same question as the previous one worded differently. Hence the students who just take Maths is 17

(3) (i) n(E) = 10

Explanation: The total number of elements in the Universal set =10

(ii) P∪Q = {a, b, c, d, e, f}

Explanation: P∪Q means all the elements in P OR Q

(iii) P ∩ Q ∩ R = {d}

Explanation: P ∩ Q ∩ R means where the three circles intersect or overlap. Clearly there is only one element that satisfies this condition

(iv) P∪Q∪R' = {a, b, c, d, e, f, m, n}

Explanation: We want all the elements in P OR Q OR in R'

(v) P' ∪ Q ∪ R = {c, d, e, f, g, h, m, n}

Explanation: We want all the elements in Q or R or not P. Remember ∪ (union) is inclusive for all of Q, all of R and all of P'

(4) Answer: (d) that is (A∪B) ∩ C = {2, 4, 6}

Explanation: Factors of 12, that is set A ={1, 2, 3, 4, 6, 12}

Prime numbers < 11, means B = {2, 3, 5, 7}

Even numbers < 8 mean C = {2,4, 6,}

Hence to find (A∪B) ∩ C, first find A∪B. We can deduce that A∪B = 1,2,3,4,5,6,7 &12, Hence (A∪B) ∩ C = {2, 4, 6}. Since the overlapping numbers between AUB and C are 2, 4, 6. Hence the correct answer is (d)

Algebra Section 14

Binomial expansion

Let us first look at Pascal's Triangle. This has the pattern shown below:

$$
\begin{array}{c}
1 \\
1 \quad 1 \\
1 \quad 2 \quad 1 \\
1 \quad 3 \quad 3 \quad 1 \\
1 \quad 4 \quad 6 \quad 4 \quad 1 \\
1 \quad 5 \quad 10 \quad 10 \quad 5 \quad 1 \\
1 \quad 6 \quad 15 \quad 20 \quad 15 \quad 6 \quad 1 \\
1 \quad 7 \quad 21 \quad 35 \quad 35 \quad 21 \quad 7 \quad 1
\end{array}
$$

Can you spot the pattern? Each number in the subsequent line is the two numbers from the above line added together!

From this we can arrive at binomial expansions.

Examples:

(1) $(1 + x)^0 = 1$
(2) $(1 + x)^1 = 1 + x$
(3) $(1 + x)^2 = 1 + 2x + x^2$
(4) $(1 + x)^3 = 1 + 3x + 3x^2 + x^3$
(5) $(1 + x)^4 = 1 + 4x + 6x^2 + 4x^3 + x^4$
(6) $(1 + x)^5 = 1 + 5x + 10x^2 + 10x^3 + 5x^4 + x^5$

You can see that the pattern follows Pascal's triangle! Thanks to Pascal a French mathematician.

However, we might need to find a much higher power of the binomial expansion.

Here is a general formula for finding the coefficient of the nth term: $\frac{n!}{(n-r)!r!}$ also written as $\binom{n}{r}$.

Factorials: (Note $n! = n \times (n-1) \times (n-2) \times (n-3)\ldots\ldots\ldots \times 2 \times 1)$. So $5! = 5 \times 4 \times 3 \times 2 \times 1 = 120$ **(the symbol! is called a 'factorial' so n! is 'n' factorial.**

Consider the binomial expansion: $(1+x)^8$. We are asked to find the coefficient of the 6th 'x' term. In order to find the coefficient of the 6th term in the expansion $(1+x)^8$ we use, $\frac{n!}{(n-r)!r!}$ where n = 8 and r = 6. Hence, $\frac{8!}{(8-6)!6!}$
$\frac{8 \times 7 \times 6 \times 5 \times 4 \times 3 \times 2 \times 1}{(2!) \times 6 \times 5 \times 4 \times 3 \times 2 \times 1} = \frac{8 \times 7}{(2)} = 28$, hence the 6th 'x' term is $28x^5$

Binomial Theorem

$(a+b)^2 = a^2 + 2ab + b^2$

$(a+b)^3 = a^3 + 3a^2b + 3ab^2 + b^3$

$(a+b)^4 = a^4 + 4a^3b + 6a^2b^2 + 4ab^3 + b^4$

Binomial Coefficients: $\binom{n}{r} = nc_r = \dfrac{n!}{(n-r)!r!}$

For example $\binom{4}{2} = \dfrac{4!}{(4-2)!2!} = \dfrac{4 \times 3 \times 2 \times 1}{2 \times 1 \times 2 \times 1} = \dfrac{4 \times 3}{2 \times 1} = 6$

Probability re-visited

Probability is defined as the likelihood of an event happening. Probability lies between 0 and 1.

A probability of 0 means that an event will definitely not happen or it is impossible to happen. Likewise a probability of 1 means is certain to happen. Probability is usually expressed as a fraction, a decimal or a percentage.

The probability of an event happening is defined as:

$$\dfrac{\textit{number of ways in which the event can happen}}{\textit{total number of outcomes}}$$

Also note that the probability of an event **not happening is 1 – the probability of an event happening**

Notation used: P(A) means probability of event A happening. Hence probability of event A not happening would be 1 – P(A).

Typical examples:

Example 1:

There are 5 red, 6 green and 7 blue beads in a bag.

(1) You pick a bead at random from the bag. What is the probability of picking a red bead? Answer $P(R) = \frac{5}{18}$
(Reason: there are 18 beads altogether, and 5 of them are red, so the chance or probability of picking a red bead is 5 in 18 or $\frac{5}{18}$)

(2) What is the probability of picking a green or blue bead? Answer $P(G \text{ or } B) = \frac{13}{18}$

Reason: there are 18 beads altogether, and the number of green and blue beads combined total 13. Hence the probability of picking a green or blue bead is 13 in 18 or $\frac{13}{18}$.

(3) What is the probability of not picking a green bead? Answer: $P(\text{not } G) = 1 - P(G) = 1 - \frac{6}{18} = \frac{12}{18}$

A simpler way of doing the same problem is to say that since there are 18 beads altogether and 6 of them are green, then this means that 12 are not green, hence the probability of not picking up a green bead is 12 in 18 that is $\frac{12}{18}$. You could of course simplify $\frac{12}{18}$ to $\frac{2}{3}$ (dividing both the top number 12 and bottom number 18, by 6)

Multiplication law in probability

When you have independent events (that is the outcome of one is not affected by the outcome of the other) then to find the probability of say event A and event B happening we simply multiply the probabilities of A and B together.

Example 1: What is the probability that we will get two sixes when a die is rolled two times?

Method: Probability that we get '6' followed by '6' = $\frac{1}{6} \times \frac{1}{6} = \frac{1}{36}$

Example 2: A fair coin is flipped three times. What is the probability it will turn up 'heads' on all three occasions? Give your answer to 2 decimal places.

Method: Probability that it turns up 'heads' **and** 'heads' **and** 'heads' = $\frac{1}{2} \times \frac{1}{2} \times \frac{1}{2} = \frac{1}{8}$ = 0.125 or 0.13 to 2 decimal places. Or you could have calculated it another way i.e. 0.5×0.5×0.5 = 0.25×0.5 =0.125 and then give your answer to two decimal places 0.13 as required.

Example 3:

If a fair die is thrown twice what is the probability of getting a 'six' followed by 'not a six'. Give your answer as a fraction.

Method: P(getting a six) = $\frac{1}{6}$, so the probability of 'not getting a six' = $1 - \frac{1}{6} = \frac{5}{6}$. Hence the probability that you get a 'six' followed by 'not a six' = $\frac{1}{6} \times \frac{5}{6} = \frac{5}{36}$

Addition law in probability: When two or more events are mutually exclusive (i.e. they cannot occur together), then the probability of A **or** B **or** C happening is simply found by adding the respective probabilities. That is p(A) + p(B) + p(C).

Summary:

(1) Probability lies between 0 and 1 and is usually expressed as a decimal, a fraction or a percentage. The probability of an event can never exceed 1.

(2) When events are independent, to find the probability of A and B occurring together we multiply the probabilities of the respective events. Remember the word 'and' is associated with '×' or multiplication.

(3) When events are mutually exclusive the probability of A or B or C happening is found by adding the individual probabilities. Remember the word 'or' is associated with '+' or addition.

(4) When working out probabilities consider whether it is 'with' or 'without' replacement

(5) You can generate tree diagrams or sample space diagrams to visualize probabilities and outcomes if it helps you.

Binomial Probability Function

The Binomial Probability function is very useful in working out the probability of r successes out of n trials. (Note in order to use the binomial probability function, all trials have to be independent and the probability of success p(s) remains constant for each trial. Like tossing a fair coin 'n' times or throwing a fair 'die' 'n' times. Or tossing an unfair coin or dice with a **known probability** 'n' times. The probability of 'r' successes in 'n' trials is given by: $\binom{n}{r} \times p(s)^r \times p(f)^{n-r}$ (where p(s) is the probability of success and p(f) is the probability of failure.

Notice that the binomial coefficients count the possible outcomes or arrangements of successes and failures.

For example I toss a fair coin 3 times. How many ways are there to arrange heads and tails if it lands 1 head?

Using the binomial coefficient model we saw earlier the number of successes and failures of getting r successes out of n tries was $\binom{n}{r}$. This means the possible arrangements of getting 'heads' and 'tails' if I throw a coin three times with one 'head' is $\binom{3}{1} = \frac{3!}{(3-1)!1!} = \frac{3 \times 2 \times 1}{(2 \times 1) \times 1} = 3$ possible arrangements.

(You can do this manually of course: HTT, THT, TTH. You can see that there are 3 possible outcomes! The binomial probability distribution simplifies this for us as the number of throws (n) becomes bigger.

Example 1: I throw a fair dice 4 times. Find the probability of rolling 2 sixes.

We know that p(s) is the probability of success is for getting a 'six' is $\frac{1}{6}$ and the probability of not getting a 'six' is $\frac{5}{6}$. Using the binomial probability distribution we can work out the probability of two 'successes' out of 'four' throws is given by

$$\binom{4}{2} \times \left(\frac{1}{6}\right)^2 \times \left(\frac{5}{6}\right)^{4-2} = \frac{4!}{(4-2)!2!} x \left(\frac{1}{6}\right)^2 x \left(\frac{5}{6}\right)^2 = 6 \times \left(\frac{1}{36}\right) x \left(\frac{25}{36}\right) =$$
$$\left(\frac{1}{6}\right) x \left(\frac{25}{36}\right) = \frac{25}{216} = 0.116 \text{ to 3 decimal places}$$

Example 2: I toss an unbiased coin 6 times. What is the probability of getting 4 heads out of 6 independent throws?

Method: Using the binomial probability distribution the probability of getting 4 heads out of 6 throws is

$$\binom{6}{4} \times \left(\frac{1}{2}\right)^4 x \left(\frac{1}{2}\right)^{6-4} = \frac{6!}{(6-4)!4!} x \left(\frac{1}{2}\right)^4 x \left(\frac{1}{2}\right)^2$$

$$= \frac{6x5x4x3x2x1}{(2x1)x(4x3x2x1)} \times \left(\frac{1}{2}\right)^4 x \left(\frac{1}{2}\right)^2 = \frac{6x5}{(2x1)} \times \left(\frac{1}{16}\right) \times \left(\frac{1}{4}\right) = 15 \times \frac{1}{64} = \frac{15}{64}$$

Example 3 18 people are given a drug for a certain illness. The probability that an individual has a side effect is 0.12.

(a) Find that no one in this group of 18 has a side effect
(b) Find the probability that in this group 1 individual has a side effect.

(a) **Method:** The probability of having a side effect is 0.12

$\implies$ Probability of not having a side effect = 0.88

$\implies$ The probability that no one in this group has a side-effect is equal to $0.88^{18} = 0.1001588…$ or **0.1002 to 4 decimal places**

(b)**Method:** Using the binomial probability function the probability that 1 individual in this group of 18 has a side-effect = $\binom{18}{1} \times 0.12^1 \times 0.88^{17} = 0.245844... =$ 0.2458 to 4 decimal places.

Example 5 A biased coin has a probability of 0.4 in landing heads when tossed. The coin is tossed 5 times. Find the probability that it lands heads at least twice.

Method: The probability that it lands heads at least twice

$= 1 - \{p(0) + p(1)\} = 1 - p(1) - p(2)$

$= 1 - \binom{5}{0} \times (0.4)^0 \times (0.6)^5 - \binom{5}{1} \times (0.4)^1 \times (0.6)^4$

$= 1 - (0.6)^5 - 5 \times 0.4 \times (0.6)^4 = 1 - 0.07776 - 0.2592$

$= \mathbf{0.6630}$ **to 4 decimal places**

Algebra Section 15

Matrices

A matrix is simply an array of numbers

As we will see matrices can be very useful for geometrical transformations such as reflections, rotations and enlargements.

Example 1: $\begin{bmatrix} 2 & 1 \\ 3 & 2 \end{bmatrix}$ This is a 2 x 2 matrix. In other words it has two rows and 2 columns

Example 2: $\begin{bmatrix} 1 \\ 2 \end{bmatrix}$ This is a 2 × 1 matrix. It has two rows and one column

So you can see that matrices (plural of matrix) can come in different number of rows and columns. <u>We need to only worry about these two types of matrices for this exam</u>

Adding and Subtracting Matrices

You can add or subtract two matrices only if they have the same number of rows and columns.

Example 1: Add matrix **A** and matrix **B** where $A = \begin{bmatrix} 2 & 1 \\ 3 & 2 \end{bmatrix}$ and $B = \begin{bmatrix} 1 & 0 \\ 2 & 1 \end{bmatrix}$

Method: Simply add the elements of each matrix as shown below:

$$A + B = \begin{bmatrix} 2 & 1 \\ 3 & 4 \end{bmatrix} + \begin{bmatrix} 1 & 0 \\ 2 & 1 \end{bmatrix} = \begin{bmatrix} 3 & 1 \\ 5 & 5 \end{bmatrix}$$

Example 2: Now consider subtracting matrix B from matrix A

$$A - B = \begin{bmatrix} 2 & 1 \\ 3 & 4 \end{bmatrix} - \begin{bmatrix} 1 & 0 \\ 2 & 1 \end{bmatrix} = \begin{bmatrix} 1 & 1 \\ 1 & 3 \end{bmatrix}$$

<u>Scalar Multiplication</u> (Multiplying a matrix by a number) is shown below:

To multiply a matrix by a single number is very easy:

For example: $3 \times \begin{bmatrix} 1 & 3 \\ 2 & 1 \end{bmatrix} = \begin{bmatrix} 3 & 9 \\ 6 & 3 \end{bmatrix}$

These are calculated as shown:
$$3 \times 1 = 3 \quad 3 \times 3 = 9$$
$$3 \times 2 = 6 \quad 3 \times 1 = 3$$

We call the number ("3" in this case) a **scalar,** and this is called "scalar multiplication".

Multiplying a Matrix by another Matrix

Example 1: Multiply matrix $A = \begin{bmatrix} 1 & 3 \\ 2 & 1 \end{bmatrix}$ with matrix $B = \begin{bmatrix} 3 & 0 \\ 1 & 2 \end{bmatrix}$

Method: $\begin{bmatrix} 1 & 3 \\ 2 & 1 \end{bmatrix} \times \begin{bmatrix} 3 & 0 \\ 1 & 2 \end{bmatrix}$ to work this out you take each number from the first row of the first matrix and multiply it by each number of the first column in the second matrix in order to get the new first element/number of the new matrix shown. You repeat this process for each row and column respectively as shown below. **It is simpler than it sounds**!

$$\begin{bmatrix} 1 & 3 \\ 2 & 1 \end{bmatrix} \times \begin{bmatrix} 3 & 0 \\ 1 & 2 \end{bmatrix} = \begin{bmatrix} 1 \times 3 + 3 \times 1 & 1 \times 0 + 3 \times 2 \\ 2 \times 3 + 1 \times 1 & 2 \times 0 + 1 \times 2 \end{bmatrix} = \begin{bmatrix} 6 & 6 \\ 7 & 2 \end{bmatrix}$$

Multiply each number in the first row of A with each number in the first column of B and add the result as shown by the horizontal and vertical lines. Repeat this process for first row and the second column and so on. **The results are shown above**.

Example 2: Given that $\begin{bmatrix} 2 & a \\ 1 & b \end{bmatrix} \times \begin{bmatrix} 1 & c \\ 2 & -1 \end{bmatrix} = \begin{bmatrix} 8 & -3 \\ 3 & -1 \end{bmatrix}$. Find the values of a, b and c

Method: Multiply the two matrices together in the usual way that is in (i) we multiply each member of the first row in the first matrix by the first column in the second matrix and we get:

(i) $2 + 2a = 8 \implies 2a = 6 \implies a = 3$

(ii) We carry on the multiplication process (1st row in the first matrix by 2nd column in the second matrix) to get

$2c - a = -3 \implies 2c - 3 = -3 \implies 2c = 6 \implies c = 3$

(iii) Finally, multiplying 2nd row of the first matrix by the first column of the second matrix we get:
$1 + 2b = 3 \implies 2b = 2, \quad b = 1 \implies$

So we find that a = 3, b = 1 and c = 3.

Just to check substituting for a, b and c let us multiply the two matrices together: $\begin{vmatrix} 2 & 3 \\ 1 & 1 \end{vmatrix} \times \begin{vmatrix} 1 & 3 \\ 2 & -1 \end{vmatrix} = \begin{vmatrix} 2+6 & 6-3 \\ 1+2 & 3-1 \end{vmatrix} = \begin{vmatrix} 8 & 3 \\ 3 & -1 \end{vmatrix}$ which is in fact the result given in the example above.

Now let us look at Identity and zero matrices.

Identity Matrix: The identity matrix I for a 2×2 matrix is simply $\begin{vmatrix} 1 & 0 \\ 0 & 1 \end{vmatrix}$

The property of the identity matrix is that if you multiply a matrix A with the identity matrix I you get A. In other words A×I = A

Proof: If $A = \begin{vmatrix} a & b \\ c & d \end{vmatrix}$ and $I = \begin{vmatrix} 1 & 0 \\ 0 & 1 \end{vmatrix}$

Then $A \times I = \begin{bmatrix} a & b \\ c & d \end{bmatrix} \times \begin{bmatrix} 1 & 0 \\ 0 & 1 \end{bmatrix} = \begin{bmatrix} a+0 & 0+b \\ c+0 & 0+d \end{bmatrix} = \begin{bmatrix} a & b \\ c & d \end{bmatrix}$

Zero Matrix

This simply when the all the numbers inside a matrix are zero. So for a 2×2 matrix the zero matrix is $\begin{bmatrix} 0 & 0 \\ 0 & 0 \end{bmatrix}$. Yes you have guessed right if you multiply a 2×2 matrix with a 2×2 zero matrix you get a <u>zero matrix</u>

Matrices can also be used for transformations of images

When we want to create a reflection image we multiply the vertex matrix of our figure with what is called a reflection matrix. The most common reflection matrices are:

A reflection in the x-axis is achieved by multiplying the appropriate co-ordinates by the matrix below:

$\begin{bmatrix} 1 & 0 \\ 0 & -1 \end{bmatrix}$

Similarly a reflection in the y-axis is achieved by:
$\begin{bmatrix} -1 & 0 \\ 0 & 1 \end{bmatrix}$

For a reflection in the origin we use: $\begin{bmatrix} -1 & 0 \\ 0 & -1 \end{bmatrix}$

(This is the same as rotating the vertices by 180°)

Finally for a reflection in the line y = x, we use
$\begin{vmatrix} 0 & 1 \\ 1 & 0 \end{vmatrix}$

<u>We can also rotate and enlarge shapes.</u>

<u>Example 1</u>: Show that P (1, 1) is rotated 90° clockwise about the origin by the matrix $\begin{vmatrix} 1 & 0 \\ 0 & -1 \end{vmatrix}$

Method: multiply $\begin{vmatrix} 1 & 0 \\ 0 & -1 \end{vmatrix}$ by $\begin{vmatrix} 1 \\ 1 \end{vmatrix}$ = $\begin{vmatrix} 1 & 0 \\ 0 & -1 \end{vmatrix} \times \begin{vmatrix} 1 \\ 1 \end{vmatrix}$ = $\begin{vmatrix} 1 \\ -1 \end{vmatrix}$

You can see in the graph below that the P(1, 1) has been rotated clockwise by 90° to Q(1,-1)

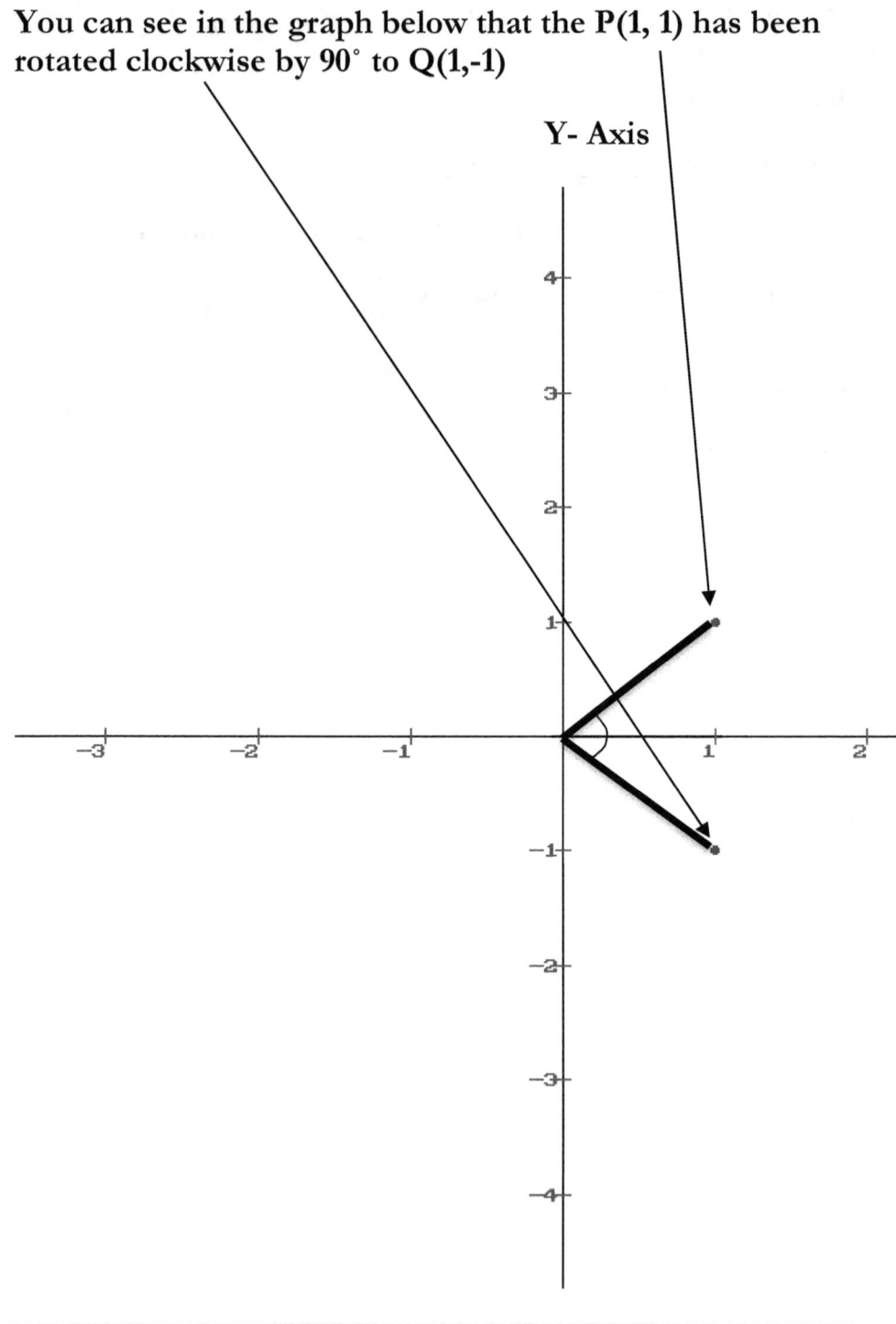

Example 2: Transforming co-ordinates by two consecutive matrices.

$P = \begin{vmatrix} 2 & 0 \\ 0 & 2 \end{vmatrix}$ and $Q = \begin{vmatrix} -1 & 0 \\ 0 & 1 \end{vmatrix}$

The point A (1, 2) is transformed by the matrix QP to A'. Find the resulting transformation A'.

Method: First work out QP. $\begin{vmatrix} -1 & 0 \\ 0 & 1 \end{vmatrix} \times \begin{vmatrix} 2 & 0 \\ 0 & 2 \end{vmatrix} =$ $\begin{vmatrix} -2 & 0 \\ 0 & 2 \end{vmatrix}$

Now multiply this by $A = \begin{vmatrix} 1 \\ 2 \end{vmatrix}$. So we get A' =

$\begin{vmatrix} -2 & 0 \\ 0 & 2 \end{vmatrix} \times \begin{vmatrix} 1 \\ 2 \end{vmatrix} = \begin{vmatrix} -2 \\ 4 \end{vmatrix}$. So the resulting transformation A' = (-2, 4)

Determinant of a Matrix

The determinant of a matrix can be found as shown below:

$$\begin{bmatrix} 4 & 7 \\ 6 & 8 \end{bmatrix}$$

The determinant of the above matrix is given by:

4×8 −7×6 = 32 − 42 = −10

What is it used for?

The determinant of a matrix can help us solve linear equations, by helping us find the inverse of a matrix, and is also useful in calculus and further math.

Symbol

The **symbol** for determinant is two vertical lines either side.

Example:

|B| means the determinant of the matrix B

(This is same symbol used for absolute value.)

Calculating the Determinant

Determinants can only be worked out for a matrix that is **square** (i.e. have the same number of rows as columns).

For a 2×2 Matrix

For a 2×2 matrix (2 rows and 2 columns):

$$\begin{bmatrix} a & b \\ c & d \end{bmatrix}$$

The determinant is simply: $|A|$ = ad – bc

Example: If B = $\begin{bmatrix} 4 & 3 \\ 5 & 7 \end{bmatrix}$, Find its determinant

$|B| = 4 \times 7 - 3 \times 5$

For a 3×3 Matrix (a square matrix)

For a 3×3 matrix (3 rows and 3 columns):

Say C = $\begin{bmatrix} a & b & c \\ d & e & f \\ g & h & i \end{bmatrix}$

The determinant for the matrix below is: $|C|$ = a(ei – fh) - b(di - fg) + c(dh - eg)

As you can see below there is a pattern to work out the determinant of a **3×3** matrix:

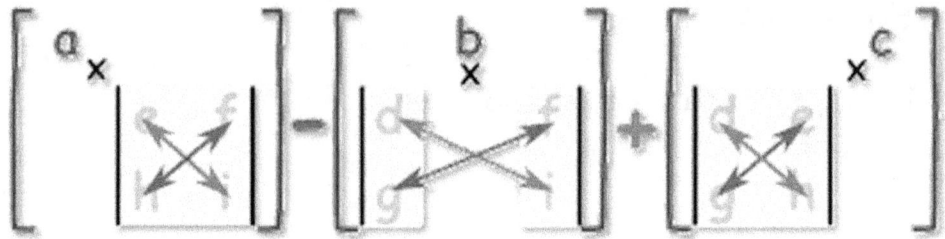

A matrices C will have an inverse C^{-1} assuming that the determinant of C is not equal to zero.

Practice Questions on Matrices

(1) If $A = \begin{bmatrix} 2 & 3 \\ -1 & 2 \end{bmatrix}$ and $B = \begin{bmatrix} 4 & -1 \\ 2 & 3 \end{bmatrix}$

Work out:

(a) AB

(b) BA

(c) 3B

(d) A^2

(2) Given that $\begin{bmatrix} 2 & x \\ -1 & 3 \end{bmatrix} \begin{bmatrix} 2 \\ 6 \end{bmatrix} = \begin{bmatrix} 10 \\ 16 \end{bmatrix}$ work out the value of x

(3) If $\begin{bmatrix} 2 & x \\ 3 & 1 \end{bmatrix} \begin{bmatrix} 1 & 2 \\ 3 & y \end{bmatrix} = \begin{bmatrix} 11 & 16 \\ z & 10 \end{bmatrix}$ Find the values of x, y and z

(4) If $M = \begin{bmatrix} 3 & 4 \\ -1 & 2 \end{bmatrix}$ show that MI = M where I is the identity matrix

(5) B(x,y) is transformed to the point B'(-1, 0) by the matrix $\begin{bmatrix} 2 & 4 \\ 1 & 1 \end{bmatrix}$. Work out the values of x and y.

(6) The co-ordinate M (1, 3) is transformed by the matrix PQ to give M'. $P = \begin{bmatrix} 2 & 0 \\ 0 & 2 \end{bmatrix}$ and $Q = \begin{bmatrix} 2 & 4 \\ 1 & 1 \end{bmatrix}$. What is the resulting co-ordinates of M' from this combined transformation PQ?

Answers to Practice Questions on Matrices

1 (a) $\begin{bmatrix} 14 & 7 \\ 0 & 7 \end{bmatrix}$

(b) $\begin{bmatrix} 5 & 10 \\ 1 & 12 \end{bmatrix}$

(c) $\begin{bmatrix} 12 & -3 \\ 6 & 9 \end{bmatrix}$

(d) $\begin{bmatrix} 1 & 12 \\ 0 & 1 \end{bmatrix}$

2 x = 1

3 x = 3, y = 4 and z = 6

4 M×I = $\begin{bmatrix} 3 & 4 \\ -1 & 2 \end{bmatrix} \times \begin{bmatrix} 1 & 0 \\ 0 & 1 \end{bmatrix} = \begin{bmatrix} 3+0 & 0+4 \\ -1+0 & 0+2 \end{bmatrix} =$ $\begin{bmatrix} 3 & 4 \\ -1 & 2 \end{bmatrix}$

5 $x = \frac{1}{2}$ and $y = -\frac{1}{2}$

6 The resulting transformed co-ordinates are x = 28 and y = 8

www.ingramcontent.com/pod-product-compliance
Lightning Source LLC
Chambersburg PA
CBHW070241190526
45169CB00001B/265